ALMOST LIKE US:
PEOPLES OF THE STONE AGE

By Ivy Hendy

Copyright@2018 Ivy Hendy

All rights reserved. This book or any portion thereof may not be reproduced or used in any manner whatsoever without the express written permission of the author except for the use of brief quotations in a book review.

Hendy, Ivy
    Almost Like Us:
    Peoples of the Stone Age
    By Ivy Hendy

I Street Press, Sacramento: 2018

ISBN: 978-1986739481

1. Stone age. 2. Culture – Origin. 3. Civilization – History.
4. Cave paintings – France.

Library of Congress Control Number: 2018943585

Cover Designed and Photographed by Jeff Hendy

...The past is the key to the present...

Contents

Preface 3

Introduction: You & I & The Stone Age Guy............................. 6

Part One: Archaic Humans... 9

Chapter 1: The Past Worlds/ Lives in the Stone Age 9
Chapter 2: Climate, Extinctions and Survival: Our Origin Ancestors 23
Chapter 3: Knuckles on the Ground: But Not All the Time 33
Chapter 4: Handy and More Human.. 43
Chapter 5: The Most Durable Human Species Ever/*Homo erectus* 61
Chapter 6: The Surprising Species of *Homo heidelbergensis*..85
Chapter 7: Who Were the Neanderthals? Or What Did Granddad See in Her? 92
Chapter 8: Is It True What They Say About Tiny People? /*Homo naledi* and *Homo floresiensis* 114
Chapter 9: More Origin Ancestors: Denisovans................... 124

Part Two: Modern Humans... 136

Chapter 10: It's A Humankind – Kind - of Thing: *Homo sapiens* 136
Chapter 11: The Climate, the Land, the Animals, the Plants . 143
Chapter 12: Out of Africa/Pathways 152
Chapter 13: Transitioning into Full-Time Human Behavior .. 161

Chapter 14: Tribalism, Rituals, and Beliefs 172
Chapter 15: No Canvas Needed/Art in Caves and Art Artifacts
.... 183
Chapter 16: Are We There Yet? 206
Acknowledgments 214
Chapter Notes 215
Bibliography and Recommended Reading 219
Index 224

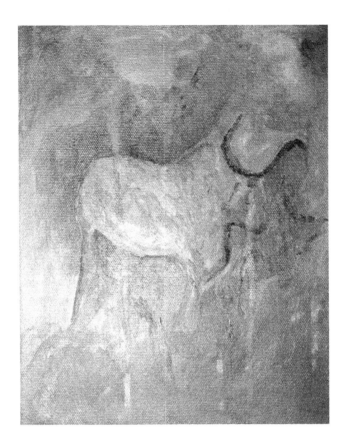

Preface

By nature, I am a lover of history. Of course, I share this love with millions of others, but it wasn't until I was 22 years old that I realized how fascinated I was with the history of prehistoric people. The year was 1964. There was a small announcement in the local paper that a man was going to give a lecture on the evolutionary story of humans. I went to that talk and eagerly sat in the audience anticipating hearing about a history of which I knew nothing. The auditorium was sparsely attended by other young college students, many reading textbooks as they waited.

The lecturer was several minutes late, perhaps owing to his age. When he shuffled onto the stage he seemed old to me. He had a full head of bright, white hair and walked with a cane. His name was Louis Leakey and though I thought him quite antiquated (he was 70) as soon as he started to speak it was clear that he was a charming man who was going to give a lively talk. He spoke about the finds he and his wife Mary had made in a country in Africa, far away from the California venue where he was that night. The country he spoke about was Tanzania where he had been raised by his British missionary parents. Leakey's accent was clear and easy to understand, but he spoke of names I had never heard before: "the Serengeti Plain", "Olduvai Gorge", "hominid" and "*Homo habilis*". He spoke of the evolution of humans from their very earliest existence on earth. And just as exciting to me, he spoke of the evidence he had found at Olduvai Gorge: stone tools; an 18-million-year-old fossil skull of the earliest known ape; and a fossil skull and bones of the first ancestor of mankind.

The experience of listening to Louis Leakey changed my perception of history and what part of it was most exciting to me. Though Leakey has passed away, I still would like to acknowledge that it was he who introduced in me an enthusiasm for the world of paleoanthropology. Though at the time I continued with my studies of philosophy, the door never shut on my interest in the field of the beginning roots of humanity.

Since then I have been fortunate enough to visit Olduvai Gorge in Tanzania and see the challenges that Louis and Mary Leakey faced. They spent three decades excavating and digging away at bare, uncompromising ancient sedimentary rock where eventually they found not only stone tools, and skulls and bones, but also footprints of the earliest humans ...saved and locked forever in the muddy floor of the gorge by an accident of drying and layering of volcanic ash. It is a sight that fills the heart with awe.

The painted caves in Spain and France are also sites that engender amazement and reverential respect. It was among the immense thrills of my life to visit not only the caves of Altamira in Spain and Lascaux in France, but many other Stone Age cathedrals, permanent testaments to the astonishing power of human creativity. The cave paintings take your breath away while speaking to us in a universal language.

I hope this book will enthuse you, the reader. It was a great pleasure to write about our ancestors of pre-history. Their story is our story. It can't be told too many times.

Introduction: You & I & The Stone Age Guy

The present time in which we live our lives circumscribes our ideas so thoroughly that when electric lights penetrate the darkness or when water flushes the commode, all this seems completely natural. But there was a time when the amenities that make modern life more comfortable were not to be seen on the planet. Life was slower, and thoughts of survival came before ideas of convenience. This was the time of the Stone Age. Archeologically speaking it is literally a stone's throw away. And it lasted a long time. The Stone Age was a transitional period when humans started to become, as if by magic, what we are today. Yet when we look more closely, there is nothing supernatural about the evolutionary journey which brought our species to the present. We and several of our close relatives managed to travel this path, though only we remain. The developmental expedition of humans can be explained as a continuous, unbroken voyage, and a venture aided by several cousins. It is worth exploring. We used to be knuckle draggers and now here we are. The question is how did that happen?

The most important item, and one that is usually overlooked on Ancestry.com, is that you and I are descended from very, very distant relatives. We'll call these ancestors "Stone Age Guy and Gal". These early predecessors don't get much press. Today Stone Age people are studied in middle school, or maybe high school. The material about them is usually curt, drab, and if you memorize the archeological terms, you'll pass the multi-choice exam.

In this book, much will be said about our archaic forerunners, the ancient ancestors whose DNA helped mold every scintilla of our very modern bodies. They are the ancestors whose genes aided to create us. This book is an attempt to bring attention to them and remember that they are not crude creatures with whom we have nothing in common. Instead, in the continuous flow of life, they are the archaic antecedents we are connected to. We'll look at where and from whom we derived. Our skin color, our smile and our brains were all works in progress several million years ago. They still are. Be prepared to share genes for future humans.

When reading about the Stone Age, the scientifically educated writers of the archeological past may discuss their subject with too much erudite verbiage, presumably for the sake of accreditation. I am not a researcher or professional historian specializing in antiquity. My hope is to write an accessible account of life in the Stone Age. I have struggled through scientific literature about this ancient era and found that the jargon of many a scientific treatise has less than a satisfactory readability quotient. My idea is to make the lives and evolution of the extinct and extant species of the Stone Age more understandable. Perhaps I should say I also have attempted to use humor and humanity in the descriptions and explanations. In technical texts the information of humans who lived, say, 100,000 years ago, is frequently bogged down in unapproachable terminology. The aim of this book is to keep things comprehensible for a non-scientific reader. Hopefully you will find it user-friendly.

If a million years of existence could be called provisional and experimental, then in this book there will be a lot of attention given to the various "provisional" million-year-old species that lived, prospered, and died out before *Homo sapiens* (us). All those helper species mean that we didn't get here without trial and error. Some of these species are still with us, under our skin, so to speak. And this means that we are not "pure" anything. Our ancestors met up with other groups who were not us, and sometimes sparks flew. The sparks are now lodged in our DNA, reminders that "purity" is just another word for arbitrarily separating people into "us" versus "them".

It is doubtful that the Stone Age Guy or Gal would harbor any hard feelings about the way we think of them. Whether we like it or not, Stone Age Guy and Gal are members of our genus *Homo*. And the irony is that archaic humans have the last laugh. If you get your DNA tested by a reputable company, you may be able to look at a chart where there is evidence of Neanderthal genes bubbling to the surface.

Another close relative are the Denisovans. They could be called our friendly East of Eden neighbors. Many of us have their DNA locked into our genetic code. When thinking of what ethnicity is made of in the modern world, the Stone Age Guy and Gal are often overlooked. We will search the past, not as students of science, but as enthusiasts of everything *Stone Age*...as in the earliest period of evidence of humans.

The start of the human past takes us back to the basics, even before stone tools. It is a lively time, nestled among the forest trees. But it couldn't last. Each era, sometimes spanning over a million years (or two or three), had to pass. Evolution is ever changing and developing like wisps of smoke, into new patterns and forms of life.

Here we are in the now...history has taken us this far. Wouldn't it be interesting to go back for a review of past mistakes and successes? Let's give it a try.

Part One: Archaic Humans

Chapter 1: The Past Worlds/ Lives in the Stone Age

Looking at time and our position in the universe, we humans can only be sure of one thing: we didn't spring, full-blown from the head of a god. Though our lives may feel demarcated, we are part of a vast continuum and a lineage whose boundaries are much more extensive than our ordinary conceptions and conceits.

For a more in-depth understanding of our place in the cosmos, there are only two directions we can look: one is to the future and the other is to the past. The ubiquitous but continually fleeting present will not help us grasp what has molded our human situation nor can the present explain the complexity that has shaped our present condition.

I have chosen to look to the long-ago past to attempt to clarify our existing knotty and convoluted present. This book is an examination of the Old Stone Age, a period when nascent ideas of human character were being created, developed, and in some sense, finalized.

So, what is the point of covering a period so distant from today? It is a time when, no matter what advancements were invented and carved out by past people, the very composition of the material they used belies the period's evanescent nature. Yet it is precisely because of the basic type of life that must have been led by Stone Age peoples and their first discoveries so long ago, that is so fascinating. The early Stone Age was on the far side of time when there were no heroes, no empires, no countries, no epic battles, no demagogic leaders, and no fabulously wealthy versus starving masses. All the drama that they experienced was about survival; communicating their thoughts was extremely limited. This is the dawning of humanity, this is when our basic character was formed, or perhaps our human character had already been shaped within the hominin design-model and as we emerged, our nature was revealed.

In any event, the humans and proto-humans who bore the burden of being so close to nature that they were ever aware of that affinity, were tested, confronted and unbowed in their efforts for subsistence. Whether they knew it or not, they were, like all the other extinct and extant animals from time immemorial, fighting for their specie's existence.

From Early Times/Not Too Different from Other Creatures

At their origin, it is obvious that the various pre-humans and embryonic humans did not see the great differentiation between themselves and other animals. They dug for tubers; they ate leaves, berries, nuts and other foliage like the many other primates that surrounded them. They made temporary shelters from vegetation, they stole eggs from nests, they searched out edible insects, and they hunted small game that they could easily catch with their hands. When they discovered fire they frequently made some fearsome mistakes. They shivered in the cold and sweated in the heat. They were part of the aggregate.

We will call our early ancestors "hominins". But we call ourselves hominins too. The term probably doesn't do any of us justice, but it is a handy means of reference for researchers and scientists to keep track of who among the history of primates were members of the human lineage. Our roots go back to African ancestors. The very earliest of our lineage chose to stay put where they came from. None of the first proto-humans ever left the African continent.

In this book we will be going back over a period connecting with our very early antecedents starting out several million years ago. The descriptions of their appearance will probably become almost unrecognizable to you...hopefully they will not disappoint you. These early hominins were furry, small bipedal individuals with longish arms. They had their day, as we all do, but during their time, starting over 3 million years ago, they were not, to put it delicately, completely human. Yet, they were hominins because they were more closely related to living people than to chimpanzees or any other primates.

Hominin history goes back millions of years; only after we come closer to the present, does history include *Homo sapiens* (us). Until a few thousand years ago, all who are in our pedigree shared the earth with other kin folks. Whether they were invited or not, there were always several close relatives, also hominins, just a stone's throw away, euphemistically speaking. Life was tough for all of them. When looked at from the perspective of ancient hominins, you can't help but feel compassion for their obvious fallible, frail position in the universe. With the hindsight that we know that many of these species went extinct without much of a trace, the whole of their lives and experiences shift, like a holographic image that changes as you tilt it.

What It Was Like

Today we often see ourselves as apart from nature; though that notion is an absurdity on its face. The modern human's ability for believing in abstract and artificed pretenses means that we build a world of conceptions we cook up in our minds. In recent times we moderns have developed intellectual models that have cleaned up the difficulty and messiness of our domain. We live in a modern world of privilege. Much of our thoughts cling to a deception that we are safe and supported with electronic wizardry which bows to our convenience. As for the rest of it, that part of life which we find unpleasant to think about, we often work to deny those aspects like death, disease, and destruction. Or at the very least, we think of them as outside our cerebral paradigm of what we believe existence should be about.

There was no such notion of separation in the Stone Age. Though food, water, procreation and protection from the elements were the dominant themes in the world of early peoples, their survival meant that they could not afford to ignore, discount or misunderstand anything. Can you imagine speaking to a person from the Stone Age and explaining to them how we buy our packaged hamburger? It isn't too difficult to imagine what a Stone Age person would think: we are not hunting our meat? Are we taking on faith that the raw ground cow flesh wrapped in an inedible covering is not rotten? Ground and not solid flesh might disappear into the fire pit as it is thrown onto the embers to be cooked. That is certainly how a Paleolithic person would see it: is it edible? (1) There was less need for subtlety. Nuance had yet to be invented. A stone was called a stone, or perhaps, at the outer limits of Stone Age imagination, a stone was called an axe.

It might seem easier to take archaic archeology on faith and devote little meaningful interest to it. Accounts of Stone Age peoples can seem too distant to be relevant. The further back in time we ask our minds to expand, the more incomprehensibly far the reach. It takes little effort to dismiss the evidence of the Paleolithic period and the people, *(our* people), who made our present world possible.

<u>Displaying and Showing Off for The Sake of Mating</u>

It is not news that most modern humans, if asked, would say that they are much more advanced in their behaviors and are in every way superior to their Stone Age ancestors. But looking more closely at the details, it's possible that our evolutionary behavior hasn't changed as much as we might think. As an example, millennia ago early humans started creating rough stone tools. These stone tools eventually culminated in one of the most important early Stone Age inventions, the handaxe. Over a million years ago our ancestors saw the value in sharpening stone and using it to increase their power. This was a groundbreaking tool that helped open nuts, seeds and shells, chop wood and build temporary shelters. Today, worldwide, the handaxe remains one of the most commonly used tools.

Although handaxes make good butchery and crushing tools and are effective for help in simple construction, the way in which the axes were made in the Stone Age did not stop at their utilitarian function. As time went on, some of the axes were etched with elaborate and exquisite designs and are found at some prehistoric sites in large piles of abundance. Exactly what these beautifully carved stone handaxes were manufactured for becomes questionable. Were these intricately fashioned handaxes used just as handy tools? The answer is that they were not. Upon examination researchers have discovered that these handaxes were never used functionally. There is no microscopic residue of ancient meat, there isn't a hint they were ever used to scrape off hides or even used to chop a piece of wood. They are, in an archeological sense, clean.

Some scientists have suggested that the handaxes produced at these Stone Age sites were not created for serviceable use in the first place. Instead, they were items used for social significance. It is thought that their manufacture was to signal attractiveness and fitness to potential mates by the person who possessed them. If this is so, the handaxes were manufactured as status symbols for males. Finding the right weapon would have had value in terms of appeal to a possible mate. Conceivably this is an example of acquiring reproductive success through exhibiting material goods. For modern humans this is an old story. But the story might be much older than we think. Some forms of display have had roots in the far-off distant past.

Learning About Us

This book is a commemoration and an appeal to remember who we are. Learning more about our primitive ancestors doesn't mean a call for humanity to collapse into bestial degeneracy. *Lord of the Flies* only covered a part of our nature. The point really is that our earliest predecessors were not a subhuman species. Deep down we are all our Stone Age ancestors; a mixture of many cravings, sensibilities, and primal urges. Without this understanding we can be caught in a trap of confusion; snared by some of our emotions, impulses and desires that roil around inside. With a remembrance that we are still inhabited in some ways by our unfettered, biological predecessors I believe we can better connect with our complex temperaments. Once grasped, this view might alter one's perception of the planet we live on and temper our dispositions toward the people who inhabit the same space and time.

Early members of the genus *Homo* who lived in the Paleolithic era chipped away at stone tools for over a million years, creating a technology that allowed them to endure, much as today we have invented technologies that allow for our own survival. Some of these million-year old archaic humans lived, gathered, and hunted both before and beside *Homo sapiens*. The next few chapters of this book are devoted to their very successful way of life. Though scientists are still counting, most members of the genus *Homo* that were not us, survived on earth far longer than the number of days of existence of our own species. Though their presence on the planet was long, the remains of their existence are still being uncovered. There will never be complete surety as to how many different types of humans existed.

Discoveries of new fossils are adding branches to our family template all the time. Our own species, *Homo sapiens*, were a part of the Paleolithic era but did not add much to technology for thousands of years. Eventually *Homo sapiens* went through a transformation, like a teenager who finally takes responsibility and starts picking up their clothes that until that time they consistently threw on the floor. That transformation in the behavior of *Homo sapiens* created the "New Stone Age" called the Neolithic period. It began, much like the period before, with stone tools. But now our ancestors discovered real estate, cereal grains, incipient religion and hierarchy. All this began ten to fifteen thousand years ago with people who developed the agricultural knowledge and skills that are used today in more than just a conceptual way. (Have you used a wheel lately?) This book will take us up to that period, 10,000 years ago, when the agricultural explosion was beginning.

This is not speculative or dicey history. The people of the Stone Age existed. Their DNA helps to form the molecular make-up of our bodies. Their primal psyches are imprinted deeply in our minds. The mistakes they made are the mistakes that we ourselves might be making. Their successes, developed over thousands of years of adaptation and experimentation, are the successes we continue to build on.

The United States has been constructed on a continent where there are few ancient cities; the great U.S.A. is less than 15 generations old. Perhaps, given its location, it is understandable that many of the citizens of the United States do not care about the lessons of prehistory. Our urban cities are not built over ancient structures, our landscape does not contain thirty-thousand-year-old painted caves, nor are there any Paleolithic digs containing Venus figurines to titillate the imagination. Few of us ever have reason to think about primeval humans living, giving birth, loving, suffering, coping, fighting, hunting, cooking, wondering about the vastness, and dying millennia ago. Taking trips to other countries, lands covered with archeological sites with remains of ancient populations which spawned present-day civilizations, might look like invented fantasy kingdoms.

Most people in the United States live in a world of sprawling malls and purpose-built fast-food restaurants. If we don't live in or near a large coastal city, and most of us do, we at least know that there is a world of modernity close by, one with glimmering skyscrapers and asphalt freeways with clogged traffic. For most of us, we never think in terms of a speculative journey back to the inconceivably long past, just to see what curiosities it might hold: no bundling up Toto and time-traveling back, clicking our red slippers and nicely asking the wizard for a ticket to the past. Living in this modernized and sculpted version of reality creates a mentality which stretches about as far back as a few centuries. A hundred thousand years is an almost mythically long time to conceive of.

Modern Research and Darwin's Theory

But whether we are aware of it or not, our mighty New World civilization exists at a specific stage and has been built within the grander sweep of time and nature. It owes its contemporary evolutionary ideas to the basic tenets of Charles Darwin's theories, and its implications to the social, behavioral and various scientific theories to the modern researchers of ancient man.

In 1859 Darwin wrote "On the Origin of Species by Means of Natural Selection, or the Preservation of Favored Races in the Struggle for Life". In the middle of the 19th century Darwin wrote: *"By the theory of natural selection all living species have been connected with the parent-species of each genus...and these parent-species...have in their turn been similarly connected with more ancient species; and so on backwards...But assuredly, if this theory be true, such have lived upon this earth."*

Darwin's work is the foundation of evolutionary biology. But there is a caveat regarding Darwin's Tree of Life. The tree that Darwin envisioned was one where the species alive today sit on the top branch and owe their existence to the ones on the bottom, the dead ones which are located by the bark, closer to the trunk. This view has now been both enhanced and rearranged by modern scientists and scientific techniques. It is no longer possible to avoid tweaking the old master's theory of evolution. Too many new discoveries have been made.

The evidence of ancient fossils is now the beginning of the examination of the Stone Age. There is a panoply of different approaches to reconstructing human prehistory. Modern forensic-style studies can be applied to fossil sites that would make Sherlock Holmes envious. CT scanning, a three-dimensional X-ray technique originally developed for medical purposes, can provide very accurate external and internal images of fossils. Lithic analysis such as microscopic evidence in bones and teeth, molecular evidence, carbon isotope ratios, chemical composition of fossils and the elements that cover them, radiometric and/or incremental dating, are all employed by scientists. The very important influence of genetic testing now allows for DNA analysis of up to 400,000-year-old bones. All these methods help to continue to advance the knowledge of our human past.

In recent years the discoveries of many fossilized bones of unknown, yet human-like species have been uncovered. With these fossil findings and other demonstrable scientific evidence, it has become clear that four million to one million years ago, and later, there were periods when our various ancestors walked the earth along with several close relatives. Darwin's vertical lineal model gives the mistaken impression that at any moment in time only one type of human inhabited the earth, and that all earlier species were older models of modern humans. But recent research has made the case that a hundred millennia ago perhaps as many as six different species of humans were meandering around the globe at the same time. It was close encounters of the ancient kind.

Most of these human and human-like relatives became extinct without leaving genetic traces in the DNA building blocks of our bodies. The presence of multiple evolutionary branches which existed together makes it difficult, or perhaps in some cases impossible, to identify direct ancestors as Darwin's original theory would have it.

Our View of History

The Greek philosopher Socrates wrote: "Wisdom begins in wonder." When we find ourselves taking a blinkered view of our place in history, we might do well to remember that our timeline stretches as far behind us as it stretches ahead of us. For proof of this, we will search across the stones, bones, fossils and dust of the Old Stone Age looking for our ancestor's roots, their habits, their actions, their successes, their failures and the cautionary tales their remnants left behind. There are many challenges that ancient people of the past solved in ways that are instructive. Some of those laudatory achievements resonate today. There are also admonitory lessons lest we dismiss humility. For instance, it is quite likely that the inhabitants of many a dust-swept ruin believed that they were the everlasting chosen ones. Of course, since they are not here, and we are, we know they were not.

Understanding and reading about the reality of the archaic past has the advantage of bringing us to a greater objectivity. What we do with this recognition hopefully will be to improve upon a plan for the sake of the far-distant days to come. The distant past actions of mankind continue to reverberate and so, too, will our actions echo into the future. The first *Homo sapiens* who populated this planet several hundred thousand years ago might not have thought of the world of the future. But that world is here. And now, or almost now, it will fade into a future world.

We are just another animal, but we are the only ones who have sufficiently evolved enough to be able to dissect and analyze our own existence. We live in a time and space very different from the Stone Age; a space imbued with scientific methods. We are fortunate enough to have that methodology to guide us and help us look back and reflect.

Questions and Perhaps Some Answers

Luckily, out of the mounds of ancient powder, fact-based information can be exhumed. From the minutiae of after-dust, remnants can produce lessons. We will sift through the information of the vestiges of our Stone Age ancestors looking for answers to questions like: Are men and/or women built for monogamy? Are we created with built-in ideas of good and bad? Is hierarchy inherent in human groupings? When was the period when females were first put into the position of the family cooks? Are we prisoners of our gender? How big a role did social affiliations play and when did such relationships first start? How malleable were hominins as far as responding to their culture? What kind of belief system did our ancient relatives create and what evidence do we have to prove this? Do the painted caves of Europe illustrate a connection to Stone Age spirit deities? In the end, the inquiry comes down to what was the identity and the influence of archaic humans?

Whenever possible, the question of how far into the past we need to travel to find early evidence of our commonality with our ancient ancestors will be looked at through the traces of evidence of early human's behavior. (2) Were the early hominins building nests or sleeping on ground level under protective layers of vegetation, indicating "house building" behavior? Why did they need to be bipedal? How do we know they were? We will look for signs of traits and behaviors that are like modern humans, searching all the way back to the evolutionary pressures that played out over prior millions of years. In the fossilized relics and remains, there are indications of a human "nature". Just like the conduct of modern humans, there is ancient behavioral evidence of aggression, violence, compassion, empathy, sympathy, competition, reciprocity, indifference, love, co-operation and altruism.

Not A Scientific Account

Taking this humanistic approach means that in this book there will be times that the stated evidence will be only somewhat verifiable. I am not attempting a *scientific* explanation of the period when humans were emerging and preliterate. Facts are facts...and there are more facts and verifications every day as scientists, at prodigious rates, uncover evidence of Stone Age populations and their life styles. Though facts are indeed facts, the explanations as to their meanings are subjective. All history is an interpretive effort. My view will differ from some but will align with others. Here I am arguing that there is a growing body of evidentiary support indicating that later Stone Age humans were not very different in either emotional or intellectual capacity from contemporary peoples. After looking at the specifics in this book, you will be the judge.

When wandering through the evidence in the fossilized records, many of the answers to the questions posed above will be considered by way of what experts in the fields of paleoarcheology, anthropology, paleoanthropology, population biology, cognitive behavioral therapy and other research fields, have provided. (*There is a bibliography in the back of this book which provides a partial list of some of the books devoted to the broad subject of the Stone Age.*) These are the researchers who have dug up, and/or analyzed the factual evidence and made their determinations. There are rock-hard remains that exist, compacted and dense, yet yielding testimony that we can ponder and interpret. Unlike asking for answers from a Providence of questionable authority or trying to divine fate in tea cup residues, the remains we will be concentrating on are remnants from our origin-ancestors.

The Characteristics of Ancient Humans

To the extent that is possible, we will explore the evolution of those characteristics that make us human—from our beginning vertical upright stance to our unsurpassed ability to cooperate. I believe that the tale that emerges helps us to identify not only our past but also our future destiny. The central point of this book is that learning from our primordial, origin ancestors will give us something solid to hang onto. Instead of closing off our vision from our remote past, it is my hope that this book will bring solace, inspiration, answers, and possible solutions as we sift through the inevitable and enduring after-dust.

(1) The Old Stone Age, or "Paleolithic era is counted as 2.5 million to 10,000 years ago. The classification is, to a large extent, based on modes of production of stone implements.

(2) *"Homo"* is the scientific classification for the genus of humans. The genus dates back over 3 million years with Homo habilis and includes all humans culminating with Homo sapiens sapiens. (Two "sapiens" appellations are awarded to our exalted species.)

"Hominid" refers to the members of the primate lineage who, along with humans, share a common but undiscovered primate ancestor. Great Apes are hominids and are related to humans, but their line diverged from the human lineage about 7 to 8 million years ago.

"Hominin" refers to pre-humans, early humans and includes modern humans (Homo sapiens) as well. Past hominins are Australopithecines, Homo habilis and Homo erectus. All these species will be covered in this book.

PALEOLITHIC AND HOMININ TIMELINE

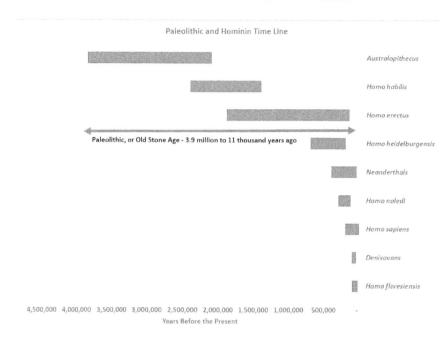

There was quite a crowd in the Paleolithic era. This chart follows the earliest of our bipedal ancestors, the Australopithecus, and moves on through the various species of the genus Homo down the path towards the first Homo sapiens. Some of these ancient humans' DNA never made it into our own DNA. But then, some did.

Chapter 2: Climate, Extinctions and Survival: Our Origin Ancestors

Hominins were born into a world of nature that was far from benign. Hostile climatic environments like droughts and ice ages, medical conditions like broken bones and infections, were all things to grimly bear. To get subsistence, most gathered plants and other edibles that were easily collected. There was constant toil for survival. With bare feet they trampled over jagged rocks, gritty sand and muck in their search for food. They needed to be keenly aware of what was poisonous and what was edible. They needed to remember paths to the best wild plant outcrops. They needed to pursue potential opportunities for sources of food without getting lost. There was always the real probability of not finding their way back to their group. The shelters they erected were flimsy and not built for permanence.

They sheltered in forests that fringed rivers or on shallow cave shelves beneath rocky overhangs. Home was a temporary abode that lasted from hours to days. Home would not protect them from dangerous predators or from a herd of stampeding animals. Cordoning off an area to erect a secure lodging to block out danger was not possible. If they had a rudimentary language, the word "safe" would have an entirely different meaning than it does in the 21st century. They were able to exert very little control. They were not masters of their fate.

If you take off your shoes and find a patch of gravel to walk on as you attempt to find a small animal to catch and kill, you might get some idea of what their daily labor was like. And be sure to go hungry. In the beginning their work and struggle was endless. It was a battle identical to the other wild animals that surrounded them. For a large portion of time in what is termed the Ice Age there was no significant dividing line between hominins and any of the other animals.

The Ice Ages

Ice ages have occurred for at least the last 2 billion years. The ice core records go back to when earth's first cooling took place. The phrase "Ice Age" is simple and evocative. But the term is not entirely accurate. Like most important things in life, there is more to the story. The distance of the earth from the sun is such that our planet can support and maintain the evolution of life. For the most part, the support of life on earth comes from the effects of our climate, a climate that allows for the presence of water in liquid form on the surface of the earth. But the earth in its orbit is not precisely oriented in space and so the amount of solar radiation from the sun that the earth's surface receives is not constant. This phenomenon causes much of the climatic changes from extreme cold to extreme drying.

Also, there are significant geological changes which affect specific weather patterns in different regions of the world. The circulation of atmosphere or water around the planet can be impeded or changed due to the dynamics of geological factors. The position of the continents and the presence of mountain ranges have a considerable effect on local geographic climate. Solar fluctuations and other geographical circumstances can alter the balance of atmospheric conditions toward a cycle of frigidity, and our earth becomes colder. This, in turn, leads to an accumulation of ice and ice sheets at the cost of liquid water. At these times the earth experiences its Ice Ages.

Thank the Interglacial Periods

There is always change over time. Climates hundreds of thousands of years ago acted similarly to the way they do now, slowly changing and morphing from one climate period to another. When the ice caps grow, they not only affect the seas and land immediately around them, but sea levels all over the world can drop by more than 300 feet because of the amount of water locked up in the ice caps.

During repeated glaciations there are also times of retrenchment, intervals of warmer periods called interglacial periods. During the last 700,000 years, cold cycles have been dominant. Climate researchers have tracked major glaciations on the earth and found that for approximately the last one million years, ice sheets have formed at roughly 100,000-year intervals. At present we are living in a brief interglacial (warm) stage.

Recently we have become worried about too much heat…. global warming is of great concern in our current interglacial period. During the past one million years interglacial stages have been the exception rather than the rule. For instance, about 90% of the last 500,000 years it has been colder on the earth than it is at present. However, this might give cold comfort about our present condition.

When thick glacial ice forms, it locks the polar regions in an icy embrace. During the last Ice Age, the higher elevations of the globe were confined to giant ice sheets. (The last Ice Age began 2.5 million years ago, but for about the last 11,500 years there has been a warming trend; technically we are still in the Ice Age.) Other, lower elevations suffered a different fate.

Africa/Untouched by Glaciers

Even in the last Ice Age there were parts of the world that were untouched by ice sheets. Such was the area of the continent of Africa. In the middle of the huge continental landscape of Africa closest to the tropics, the Ice Age caused a process of desertification. This is also the approximate region in Africa where the first of our human ancestors arose. None of our cousins in Africa were walking on ice cubes, but instead they were shuffling through a changing landscape of dry grasslands as vast surfaces in the middle of continental Africa were slowly undergoing drought and the consequent decline of forest lands.

The Hominins 3 Million Years Ago

Because from time to time there were periods of extreme cold on our planet, the survival of the various forms of the genus *Homo* was always problematical. Three million years ago—around the time the first hominin species appeared—Africa was switching from wooded areas to open grasslands as the climate dried out. Though in other parts of the world ice sheets were spreading over 30% of the globe's surface, the African landscape was changing to a drier and less forested landscape.

Humans 250,000 Years Ago/Ice and Fire

As for our own species, *Homo sapiens,* 250,000 years ago when we debuted in Africa we nearly went extinct. One of the times of the earth's many maximum glacial activities corresponds to the approximate time of the first *Homo sapiens*. We were originally welcomed to a world of ice. We survived by using fire.

As we will see later, the more successful early humans sought diverse food options during variable periods of climate change, even as the African landscape was slowly trending toward a more uniform grassland environment.

Animals That Were Equipped to Be Warm and Toasty

Three million years ago, an era when the first hominins (bipedal apes) appeared, was also the time when extraordinarily large animals roamed the savannahs of Africa. These giant animals were called megafauna. (1) Their bodies had adapted to combat the cold by growing insulating coats and blubber. When the land was one big freezer storage compartment, the Ice Age might have frozen the paws off any vulnerable mammal that roamed the landscape without layers of protective fur and fat.

Through hundreds of thousands of years, as the climate continued to change, these giant beasts did not survive. Archaic humans were living alongside these beasts. It is not possible to say if any form of the genus *Homo* influenced the extinction of Africa's megafauna. Africa's megafaunal extinction occurred around 1.4 million years ago.

Below is a partial list of some of the giant animals, now extinct, that existed on the African savannah as recently as l.4 million years ago:

- Giant, lion sized versions of hyenas;
- Tall, blade-toothed dogs;
- A saber-toothed cat called *Dinofelis*. This cat roamed the African savannah from 5 to 1.2 million years ago. It was an ambush predator that needed to stalk within close range, taking its prey by surprise;
- A member of the elephant family, called *Deinotherium,* which was the largest land mammal on the planet at that time. It lived between 20 and 1.2 million years ago. It stood 14 feet tall, far bigger than any living elephant and weighed over two thousand pounds;
- A giant giraffe called a *Sivatherium* which possessed two large deer-like antlers;
- The *Ancylotherium,* which went extinct 1.4 million years ago, had short hindquarters and powerful arms with long claws. It had a body slightly resembling a gigantic elephant with the head of a deformed donkey. It is thought that it bent its forepaws in a way that caused it to walk on its knuckles like a gorilla;
- Giant versions of warthogs and a massive version of wildebeest, plus a colossal-sized zebra species, all now extinct.

These megafaunas existed when our archaic ancestors were roaming the savannah.

Though none of these species have survived, the genus *Homo* did, as witnessed by the fact that we are here today. All these giant creatures plus many more disappeared around the time when one of our early ancestors *(Homo erectus* 1.4 million years ago) was developing their stone tool technology and experimenting with fire.

So-long, Adieu, Cheerio

The effect of glaciation was not the only factor that caused the disappearance of the large land animals (and some of the species of the genus *Homo*) in Africa. Mass extinctions of many of the ancient animals might have been the result of several factors: asteroid impacts; the somewhat erratic wobble of the earth; tectonic plates that groan, shift and realign; the changing of the earth's ocean patterns; enormous exploding volcanoes - there have been many theories about the causes of mass extinctions of the past.

A Story of Endurance

Our story is about the history of the Old Stone Age and that means the time when our hominin ancestors and humans come into the picture. They first arose in Africa and then traveled elsewhere in the world. During the various advances and retreats of the chilly Ice Age, tiny groups of a variety of intrepid archaic humans and much later, *Homo sapiens*, worked out how to survive no matter how harsh the conditions. These various hominins were few and for a very long time, their impact on the world around them was minimal.

Traveling or What Scientist Call "Dispersal"

Even if you have a grandma who never left the town she was born in, our species, by nature, are travelin' folks. Two hundred thousand years ago, when it was cold, a majority of the earth's water was frozen into ice. There is a favorable side to much of the waters in the rivers and seas being frozen over, locking moisture into spreading glaciers. When sea levels drop, land passages that were underwater and inaccessible before, become available as corridors to reach new continents. Several hominins who lived in Africa decided to strike out for different lands. Archaic *Homo sapiens* began leaving Africa in very small groups as early as 300,000 years ago. *Homo erectus,* a close relative, said good-bye much earlier than that.

Adaptation in The Age of Climate Change

In all times on planet earth, evolution has a way of challenging each species. Sweeping changes cause natural adjustments and new ways to grapple for existence. Ultimately, animal species and plant species are either going to adapt and survive or perish. The more flexible the species is to broad forms of adaptation, the more likely their evolutionary survival.

Survival of the Hominins One Million Years Ago

A million years ago in Africa, large land animals that had roamed for eons had fewer and fewer offspring. Creatures that had huge body masses and fat reserves were disappearing. Yet, some of the small but indomitable creatures of the genus *Homo,* survived.

As the hominins increased in population, they started making their habitats in the same areas that had been inhabited by the disappearing species. The hominins were also developing new adaptation techniques. As early as almost three million years ago, occasional stone tools were probably being created. By one million years ago, stone technology had incrementally progressed.

A popular theory among some researchers is that overhunting by our archaic ancestors was, in part, responsible for the extinction of the large cold-weather predator and prey animals. Though this is postulated by some scientists, the evidence has not been firmly established since there were very few hominins in existence a million years ago. Researchers estimate that a million years ago there were less than 20,000 breeding individuals; less than the population of gorillas today. This means that archaic humans were an endangered species

The Change to the Holocene 11,700 Years Ago

Changes that dramatically modify food supply and climate were in the air about 12,000 years ago. The climate in Africa began to warm. It is called the Holocene which is the name given to the last 11,700 years of the earth's history. It denotes the time since the end of the last major cold epoch. Since then there have been small-scale climate shifts, but in general, the Holocene has been a warm, interglacial period.

For thousands of years our own species, *Homo sapiens*, had dispersed all over the world, so that by the beginning of the Holocene period humans were present on most continents. Adaptation, traveling, and adjusting to new surroundings was part of the indomitable nature of the lineage of *Homo*.

And adjust they did.

The Stone Age: A Human Story

There are many fascinating historical adaptation techniques and features that define each hominin species. If researchers of the future analyze our present *Homo sapien* cultures, they will look at our food sources, social hierarchies, cooking and food preparation, agricultural management, technological skills, belief systems, and adjustment to habitats. These are the same things that are important aspects and considerations for Stone Age humans.

The Old Stone Age is defined by archaeologists as a broad prehistoric period (roughly 2.5 million years ago to about 10,000 years ago), a period during which our ancestors created stone implements which they developed from the rudimentary technological skill of pounding cobblestones into an edge or a point and eventually creating many other devices made of stone. The period of the Old Stone Age encompasses the many hominins (proto-humans, archaic humans and humans) that used these stone implements.

It is important to understand that looking at stone tools of the Stone Age is a way that researchers categorize and date the existence of early hominins. It doesn't mean that only their stone contrivances are meaningful. But it does mean that when looking so far back in time, anything in stone will have endured for a much longer time than implements and artifacts made from material that is less durable.

Hunter-Gatherers

Timing might be everything, but the Stone Age records are approximate and vary depending on where the locations of the fossil sites are situated. Hunting and gathering has been the method and lifestyle of most humans who have ever lived. The upside in the Stone Age was that hunter-gatherers could remain nomadic and not be burdened with property rights and responsibilities. In the bigger picture of human history, our modern method of cultivating the land and growing food crops is a brief aberration.

But hunting and gathering is a less effective way to manage food consumption for large populations. This meant that all Stone Age populations were small. Before 10,000 years ago, the human population of the entire planet at its zenith was smaller than that of today's London.

Most of the hunter-gatherers in the Stone Age existed by grouping together in small, nomadic bands that foraged for edible plants and scavenged for pieces from predator-killed animals. As time went on, they assisted each other for hunting purposes. When times were spare, sometimes the fruits and roots from their gathering activities were all that was on the plate. At other times they found they could both hunt and gather. Fortune always plays its hand. Luck should be one of the component elements listed on the Periodic Table.

Before we leave the after-dust of very ancient hominins, let's look at some of the more celebrated.

> *(1)* *Technically, Megafauna is any animal over 100 pounds, which includes us.*

Chapter 3: Knuckles on the Ground: But Not All the Time

Planet of the *Australopithecines*

The species of hominin that is called *Homo sapiens* is not actually better adapted than any late-great and extinct cousins of ours. Instead, the latter were just as well adapted, until various environmental and climatic conditions changed sufficiently to do them in.

Planet of the Apes is not a true story. But if it were, it would paint a more engaging picture since we could easily visualize a common ancestor who walks in a bipedal position, is about our height, and wears cool Hollywood inspired uniforms, (and of course, speaks English). As it is, if you go to the zoo and look in the chimp and bonobo cages, you'll see our close ancestors whom we branched off from roughly seven million years ago. Still, you can be heartened that we come from a common ancestor, not directly from chimpanzees and bonobos. The joint ancestor of chimps and humans who lived well before seven million years ago used to be called "the missing link". But that is yesterday's news. Now is it fashionable to talk about DNA. Humans share 98.6% of their DNA with chimpanzees. This might explain the reason why the average person in the United States consumes twenty-five pounds of bananas a year. Just saying.

Sharing Primate Characteristics: Chimpanzees and Humans

What makes us human? Before you decide that this question has an obvious answer it should be mentioned that we share a lot of characteristics with other primates, particularly the Bonobos and Chimpanzees. Besides an expanded brain, we share other physical traits. "The snout is out" might be one ringing refrain when it comes to chimp/human similarities. And binocular vision, with the eyes close together and directed forward so that we see one image with our two eyes is also an important primate trademark. Overlapping vision happens when the eyes are located on the front of the head, not on the sides, over or under our ears. (What a relief.) The overlap allows each eye to view objects with a slightly different, focused viewpoint. Luckily for the artists among the primates, we can also discriminate among different colors. The claws vanished from primate's digits in favor of nails. Having fingernails on the ends of fingers rather than claws allows for the full development of the grasping hand. Also, typically most primates have only one offspring in each pregnancy. Before the invention of EvenFlo Baby Bottles, think of how difficult it would be to feed six babies with only two mammary glands. Yikes!

Locomotion Moves the Knuckles

A walking technique that chimpanzees do not share with humans is their form of locomotion. If you watch chimps, they have an ambling gait that includes walking with their two back legs and most of the time they use their two arms as well. Knuckle-walking is a form of movement where the forelimbs hold the curved fingers in a partially flexed position. This allows a chimp's body weight to press down on the ground through the knuckles. With each step, the knuckles and the back feet contact the terrain.

Australopithecines (plural), were the first mammals whose ancestry led to the genus *Homo.* They had many of the same characteristics as chimps. But as they prowled around their homeland in the African savannah, their gait was decidedly upright.

Use Those Two Legs and Let Your Arms Go Free

There are theories about why standing upright and walking on two legs is a good idea. For one thing, standing on two legs lets you see more stuff. We have all seen pictures of grizzly bears standing upright to look around. And we have seen squirrels stand upright for the same reason: to get a better view, possibly of the grizzly bears. For the most part, the *Australopithecines* didn't bother about a four-legged stance. They went for the jitterbug rather than the bunny hop.

And they were right. Walking on two legs, with the femurs located directly under the hips is the most efficient way to get around. When chimpanzees are tested on a treadmill, they require 25% more energy than the two-legged human variety.

I Love Lucy

Australopithecus (singular) or Southern Ape has to its credit the prehistoric recovered skeleton that is named "Lucy". [1] She and her folks are renowned as being the earliest pre-humans. *Australopithecines* lived in Africa 7 to 2 million years ago. They were not as given to walking in only upright positions as later hominins. In fact, though *Australopithecines* are placed in the first chapter of anthropology textbooks, they can't really be said to be human, though they are hominins.

Above for Safety/Below for Foraging

Between us and the *Australopithecines*, there are 100,000, maybe 200,000 generations. Though adapted to the African environment in which they lived, these short (they were no more than 4 feet tall) creatures created living arrangements that would stymie a modern human.

It is thought that Lucy and her family lived in trees by night and climbed down to go browsing for food by day. She and her kin had arms that were much more elongated than yours or mine. With a sloping forehead, a pressed-in flat nose, and a weak lower jaw that set out her prominent bicuspids and canines, her physical characteristics were not ones that *Homo sapiens* later kept as their distinguishing features. A chimpanzee would perhaps recognize Lucy as a distant cousin, if they were to bump into each other while lapping up water on a lake in the savannah. If Lucy lived today, we would probably put her in a zoo or a nature park. We would not give her legal rights to drive a car or invite her for dinner.

What Big Jaws You Have, or Maybe Not

While searching for fine dining, termites and insects were eaten, as well as an occasional piece of raw meat. There is evidence that Lucy and her group cut the meat with stones they would come upon, but only in an opportune manner, not as a rule. The jaws of many of the *Australopithecus* family show that their tastes were generally inclined towards vegetarianism. Their basic foods were plants: fruits, seeds, roots, tubers, bark, nuts and fruit pits.

Compared to other apes, all humans and proto-humans have weak jaw muscles. Even as far back as 3 million years ago, *Australopithecines* had flimsy jaws. Their jaws were not made for tearing into live game animals in a ruthless rush of predator brutality. But though their jaws were not as strong as their great ape competitors, the molars of *Australopithecines* were sturdy, and this meant they could chow down large quantities of pliable food in a relatively short time. There is an advantage to eating more food in less time without the necessity of spending lengthy intervals chewing. In case you haven't guessed: the less time spent with heads down, concentrating on eating, and bottoms pointing upward in a vulnerable position, the better. It's a rule to remember the next time you go out to eat.

Large Amounts of Salad: Hold the Dressing

Since most of the diet of *Australopithecus* was based on plant carbohydrates, they could only sustain vibrant energy if they ate great quantities of vegetation. Much more material must be eaten when surviving on a plant-based diet than on a diet that is based primarily on raw meat.

Survival for *Australopithecus* was challenging. The land was a vast sea of animals and many of the bigger ones, both the ones with four legs and the ones with large wings, hunted the smaller proto-humans as tasty tidbits. Because they were little and didn't have the brute strength of the heftier predators, *Australopithecines* were often the prey that was served up on the dinner platter.

Steak Tartar and Hold the Horseradish

Researchers have found evidence that *Australopithecus*, when they could, ate raw meat as well; meat that came from the carcasses of animals that stronger and higher-order predators had caught, killed and then left partially eaten and discarded. Not that this activity comes from our primal relatives, but it must be mentioned that left-overs are still a favorite today.

Parasites and Our Ancestral Hominins

Parasites, abundant and ubiquitous throughout evolutionary history, were around long before the dinosaurs. By the time *Australopithecus* appeared on the scene, the number and varieties of parasites living in the guts of other wild animals and waiting to make their home in the guts of a proto-human were too numerous to count. Eating animal flesh gave *Australopithecus* more calories in a concentrated form, but there was always the chance that they could contract parasites from eating raw, infected animals. And the effect could be fatal.

Hands and Feet but Not Much for Manicures

Australopithecines like Lucy had appendages we would recognize. She had reasons for walking upright, reasons that included many advantages: she and her group could carry food with their arms and hands. But though Lucy could gather things with her grasping hands, her hand structure was different than ours. She had long fingers but short thumbs. If you fold your thumb over and try to touch it with your index finger you will find it is awkward and clumsy to pick up small, delicate items. But long fingers served a day-land walker/night-tree dweller very well. Who needs to be adept at cell phone texting and knife and fork etiquette when swinging from branch to branch?

Of course, to climb up those trees, well adapted feet would be necessary. *Australopithecines* had opposable big toes which looked and acted like our thumbs. These useful toes were appropriately suited for climbing and for hanging onto hazardous woody boles and swaying tree limbs.

<u>The Social Club</u>

From all evidence, *Australopithecines* were social animals. Like monkeys and apes, they quickly began living in groups. But social animals form hierarchies and so Lucy, like all her kind, had to navigate the complex web of friendships and rivalries. We will never know if Lucy was the Queen of the May or the wallflower that swung from the lower branches. Because of studies of monkeys and great apes, we can be sure she had a niche to fill in the social pecking order, whether she liked it or not.

<u>In Our Sleep: The Embryonic Call of Danger</u>

There are many ways in which we are connected to our primitive ancestors. Perhaps we can feel their presence in our sleep. It is premised that just as you are falling asleep and you are dozing off, you might be awakened by a jerk of your leg or maybe your entire body. That hip-jerk from sleep will bring you to full attention. *Australopithecines* walked and foraged by day, but at night, when they climbed up in a tree for sleep, they still had to be vigilantly alert for danger. As they were dozing off to sleep, their still active brains might tell them, "Beware! Danger!

Falling!" Their hip-jerk reaction would take over to abruptly awaken them. None of them wanted to be torn to pieces by the saber-tooth cats prowling around on the grounds below.

So, too, it is theorized that the hiccup in our own sleeping pattern is the result of an arousal of our primate brains and for a moment, we hear the prehistoric warning of primordial danger. After all, as the theory goes, there is no reason for our modern brains to worry about falling off the mattress. It is an *Australopithecus*-sort-of-feeling...even if it isn't very warm and fuzzy. But it is doubtful that any of the deeply primitive instincts within us are evocative, reassuring responses.

Australopithecus: Several Hat Sizes Down in Comparison

So, what about brain size? Lucy and her group had brain sizes of about the dimension of a modern chimpanzee's. Researchers are quick to point out that relative to her body size, Lucy's brain was quite hefty. Nevertheless, the many hominin species to come after *Australopithecines* grew much larger brains.

Brain sizes are connected to an upright stance. As humans walked upright, they grew ever stronger vertebrae, and those stronger spines were able to hold a heavier head weight. If you remember your mother's admonition to "stand up straight" you will see she had good reason for her concern. Intuitively, she must have known that good posture was connected to bigger brains. (Mothers know these kinds of things.)

Yet on an evolutionary scale, the connection of a straight, vertical stance and a larger brain size was not obvious. The structural changes that occurred in the size of the brain and many of the modifications in hands and feet didn't happen during Lucy's family's several million-year existences. And the lack of a larger brain might be the reason that *Australopithecus* did not appear to be remarkably different from the untold numbers of other organisms on planet earth.

Things That Go Bump in the Night

Just because we have larger brains than the smaller, more ape-like *Australopithecus* doesn't mean that certain characteristics weren't passed on. Modern humans have an innate tendency to be frightened by a loud roar. Fright-night movies play on our inborn instinct to jump out of our movie seat, ready to take-off when an ominous loud howl suddenly shatters the air in the dark cinema theater. The reaction to be afraid of loud and strange noises might well be instilled into our brains from some distant instinct to wake-up and pay attention to danger. Beware! The cave bear is close-by. Run!

Why Can't a Guy Stick Around?

The *Australopithecus* males were much larger than the females. An *Australopithecus* female averaged about sixty-six pounds while the males, weighing in at over one hundred pounds, were almost half as large again as their female counterparts. In contrast, modern human males are only about 25% heavier than modern women. When looking at behavior patterns of today's chimpanzees and great apes, the large size of the male *Australopithecus* suggests that they were polygamous; with a few dominant males monopolizing the troop of females.

Social behavior doesn't fossilize, but it is thought that three million years ago Lucy and her group had a chimp-like social structure. In most mammals, the females stay in and around their group's chosen territory, and after adolescence the males disperse to other areas, presumably to avoid inbreeding with their natal group. They instinctively find another group of females to mate with outside their direct lineage. Then, after the ball is over, the males once again leave the gals in the dust.

This seems to be the case for *Australopithecus*. The evidence comes from the record of fossil teeth showing *Australopithecus* male dispersals. Putting the fossil record together with studies of modern chimpanzees, it is thought by many researchers that male *Australopithecines* fought for hierarchy. When the pecking order was established, there was promiscuous, multi-sex mating. Not to speak badly of our ancient ancestors, but there seems to be evidence of intraspecies killing and even cannibalism among some members of this species. Certainly, there could have been aggressive behavior between bands of neighboring males that breached the troop during the brief mating season. Natural selection is not necessarily a benign force.

But wandering off and scattering to other areas leaves little time for bonding and paternal protection of offspring. It was not, apparently, during the days of Lucy that pair bonding arose. Like chimpanzees, it is doubtful if individual male *Australopithecus* could recognize their offspring.

Sex and the Single Girl

Though pair bonding and its emotional implications evidently did not exist as it does in modern humans, female *Australopithecines* were probably not entirely without control over who they mated with. Females of most species get quite particular when it comes to whose male DNA will help to create their offspring. This is the reason males of many species display cumbersome antlers or unwieldy tail feathers. There is such a thing as female power, it is just sometimes subtler.

Genes, it turns out, are relevant to everything. And the beauty of their power is that no one need know a thing about DNA for it to be the controlling factor in conduct. Much of the behavioral factors of hominins were created in their genome: their chromosomes; the sequences of DNA that were destined to be duplicated in every cell of the species. The role genes played in the behavior of the earliest pre-humans is still being debated.

The female *Australopithecus* might very well have had a say in choosing her mate by noting features like the male's strength, his enthusiastic holler, and size of his canine teeth (what else?). A large body size for males is the most common feature of selection in the animal world. And as we have seen, the male *Australopithecus* was a giant compared to his female counterpart.

Just as in the comportment of other mammal species, the female *Australopithecus* had to be choosy about what genes her infant would wind up with. She must have had a strong influence because the species of *Australopithecines* lasted for several million years. Perhaps modern human females have this same discriminatory choice of distinctive partner selection imbedded in the primordial corner of their brains. But if your mother doesn't think you chose a sensible partner, she might say you forgot to check-in with your elemental brain cells.

> (1) Several anatomically somewhat different types of Australopithecines existed at this time: the two most often cited are the gracile or lighter-built types like Lucy and her group, and the heavier forms called robust. The robust Australopithecus are sometimes placed in the genus Paranthropus. Robust and gracile ate different diets and lived in different areas in Africa. Their dietary difference helps to explain their different anatomies.

Chapter 4: Handy and More Human

True Love and Handy Too: *Homo habilis*

Are there any advantages to males keeping a close association with their mates, so close that in the morning they argue over who is going to make the coffee? Yes, is the simple answer. Single pair bonding means that the male will help guard the female. Further, they might recognize their own paternity. In this way the males give special attention to their off-spring. Juveniles can take longer to develop, and they have the companionship of daddy as well as mommy to help protect and feed them.

Olduvai Gorge: Where the Living is Easy

This takes us to another hominin, one who had more interest in sticking closer to his mate and children: *Homo habilis* or Handy Man. *Homo habilis* is one of the most famous of the fossil species. These fossils were first discovered in 1960 at Olduvai Gorge in East Africa by Louis and Mary Leakey. In case you missed the news, this fossil species was originally touted as the link between *Australopithecus* and us (*Homo sapiens*).

There are some signs that *Homo habilis* was closely linked in their pair bonding relationships. There are fascinating fossils of *Homo habilis* footprints in the East African river beds of Olduvai Gorge: small footprints side by side with larger footprints fossilized together, as if perhaps the bodies that created these footprints were used to walking together and had a familiar and or even an intimate relationship.

What's Love Got to do With It?

Did *Homo habilis*, these early hominins of ours, have yearnings other than the urge to eat and procreate? Was there, for instance, a dreamy infatuation in the eyes of the pair bonding mating couples? Something must have kept them in their semi-permanent, exclusive relationships. The fossil footprint evidence is one reason why it is theorized that *Homo habilis* males and females and their families stayed in pair connections and were also part of a larger group.

If love as we know it was not the glue that kept the couple together, then perhaps it was an unconditional commitment to the individuals in the group. For survival, each member of the group needed the other members for their continued existence.

The First Humans

Since paleoanthropologists are discovering more and more varied fossil finds, a literal rock-solid answer as to which human species came first is no longer possible. But it is safe to say that *Homo habilis* lived 2.3 million to 1.4 million years ago and was among the first human species to exist. Researchers know something about them because of the geography of the savannah in East Africa. The Great Rift Valley volcanoes, where the Leakeys found the first fossils of *Homo habilis*, deposit volcanic ash in predictable layers so that dating fossil finds is particularly convenient.

Brains but Not Brainy

The modern human brain is about four times the size of an *Australopithecus* brain. Big heads are part of our charm, even if the saying about having a big head is a criticism. On the list of "What makes us human?" after a bipedal stance, increased brain size is one of the fundamental definitions.

Homo habilis' brain case was fifty percent larger than *Australopithecus*, but they still retained some ape-like features including the long arms. If their features didn't rival a more modern human's, they did grow a bit in weight and stature, weighing in at 75 to 121 pounds and standing, up to 4 feet 11 inches. Their average lifespan was 18 to 20 years. With such a short life, they grew up, but they never grew old.

Though *Homo habilis* had a bigger brain than *Australopithecus*, compared to modern humans, *Homo habilis'* brain capacity was fifty percent smaller. Yet, if a haberdasher was asked to show Handy Man some hats, he would have to up-the-ante if he only specialized in fitting *Australopithecines* for jaunty caps.

At one-point, Handy Man shared the earth with *Australopithecines*. The two-species existed in various parts of Africa two million years ago; one of the species was going out-of-date (*Australopithecus*) while the other was just beginning (*Homo habilis*). It is possible that they met each other at their local Stone Age clothing store. If so, what they thought of one another will have to be left to the speculation of history-- though one would hope that no gloating over larger and bigger heads occurred.

What Big Tools You Have Mr. Handy Man

At this juncture in the story of humanity we are not quite at the period when it is appropriate to use the accepted definition of the word "culture" for *Homo habilis*. In the rough and tumble life they were faced with, just the task of keeping their children alive and buffered from the menacing world was a triumph.

Although there is one aspect of *Homo habilis* that shows an important step in the advancement of human evolution. This step can be interpreted in a broad sense as a move toward a culture: *Homo habilis* is credited with developing stone tools. In fact, their tool making abilities define *Homo habilis*---which is why *Homo habilis* is nicknamed Handy Man.

What Handy Man was doing with those tools was modifying and altering his physical environment. Their patience was apparently at an end: no more were they entirely at the whim of nature.

Their new motto was going to be about looking out for oneself with homemade implements. You might look at this innovative new outlook as the beginning of pragmatism.

Stone Tools Were More Than A Hobby

Nor was it just one fossil of Handy Man with one stone implement that has been discovered. Frequently stone tools are found with the fossilized remains of this species. (1) Chipping away at stones used as instruments to get at meat must have been thought of as their primary work. There is evidence of their accomplishments in many different places in Africa. Find fossilized bones of *Homo habilis* and it is certain there will be stone tools as well. It must have been a skill they were eager to carry out. And of course, they taught that skill to their young. In this sense, if we take the broad view of culture as learning that takes place socially and is passed down, then *Homo habilis* may be said to have had a rudimentary form of culture.

In banging and flaking away at stone, Handy Man was only thinking of butchering animal flesh. Creating a spear with a handle and hafting on a pointed stone that could be used as a weapon for killing prey was beyond his pay grade. Instead, during the over one million years of time that *Homo habilis* existed, the revolutionary technology of creating stone tools for scraping the meat off the bone never developed into a thriving business in blade making. Projectile pointed stones were to be left to Handy Man's future cousins to invent.

Smart, But Not Smart Enough/a Meal Fit for a Cat

With a bigger brain, the intelligence and social organization of *Homo habilis* was more sophisticated than *Australopithecus*. Yet, like the *Australopithecines, Homo habilis* used stone tools for scavenging; cleaving meat off the bones of discarded wildlife.

Despite more sophisticated tool usage, there is ample fossil evidence to show that Handy Man was a popular staple in the diet of large predatory animals. More than one of the fossilized bones of Handy Man has turned up in the stomach of creatures such as the now-extinct large toothed carnivorous cat called *Dinofelis*...also known as a 2-million-year-old giant jaguar.

When Handy Man Was Not the Meal on the Menu, What *Was* on the Menu?

Because of the use of stone tools, *Homo habilis* was not as interested in an all- vegetarian diet. Handy Man is the first recorded day-to-day omnivore in the *Homo* lineage. How tasty Handy Man's three course meal was, is not recorded. But the number of exotic animals to choose from was greater than what is offered at Whole Foods. And what better way to use a stone tool, besides carving off the flesh, than to smash open the bones and suck out the nutritious marrow? That was the culinary practice of this species.

Foraging: A Search for Haute Cuisine

In many paleoarcheology textbooks, *Homo habilis* is defined by the way they acquired their food. Since the invention of stone tools was primarily used to obtain animal flesh and aid in gathering wild plants, it seems a noteworthy and significant trait.

For the bulk of human history our ancestors were hunters and gatherers, (also known as foragers). Hunter-gatherers don't tend crops or herd animals. Their foods are plants, game meat, fish and insects.

When *Homo habilis* went about gathering wild plants and killing or scavenging edible animals, they were faced with two possible eating practices. You've already guessed that it was not whether to use linen or paper. The first response is for the hunter-gatherers to consume the plants and prey directly on the spot.

Though morals and/or etiquette don't fossilize, let's generously believe that the idea of sharing was already an ingrained response, possibly helped along by larger members of the group demanding first dibs. The other technique is to take the bounty back to the campsite and store it for varying lengths of time. Presumably *Homo habilis* and their group did much more of the former than the latter. Yet many modern animals, including wild cats and birds store/hide their food. It would not be beyond reason to think that *Homo habilis* did so as well.

Another characteristic of Handy Man and all hunter-gatherer populations who came after them is that their social life would have centered on their mobility. A high degree of mobility is a necessity for foragers who need to move around their environment in search of food. This is still true with the few remaining nomadic tribes living in the present day

The Long Habit of Foraging

The lifestyle of hunter-gatherers represents a long phase in human evolution. *Homo habilis'* method of foraging to obtain sustenance epitomizes the existence that engrossed most of the waking hours of all of our early ancestors. It wasn't until approximately 10,000 years ago that human lifestyles changed from mobile hunter-gatherer behaviors to sedentary agrarian practices. We are still omnivores, but our methods for obtaining our dinners have changed significantly.

Possibly because of *Homo habilis'* early tool innovation and a diet that was not specialized and which included whatever was edible and available, this species of hominin became the precursor of an entire line of new species, culminating in us.

Population Boom

Researchers have found that Africa was a busy place during the era of Handy Man. Since the 1960s when *Homo habilis* fossils were discovered, many even earlier hominin fossils have been uncovered which now have created debates within the paleoanthropological community as to which hominins came first, and which hominins belong to what species: did this or that fossil hominin skull belong to the group of proto-humans like *Australopithecus*, or to the more human-like species such as Handy Man? Since many of these debates among researchers have been going on for decades, the definitive answers might come after we ourselves have fossilized.

<u>*Homo rudolfensis*/Smarter Than the Average Hominin</u>

One of these "new" fossil finds is *Homo rudolfensis*, (1.9 to 1.8 million years ago), who lived in what is now northern Kenya near Lake Rudolf (aka Lake Turkana). The *Homo rudolfensis* skull shows a larger brain size than *Homo habilis*. What is particularly significant about *Homo rudolfensis* is that its brain grew and expanded into the front region of the skull called the frontal lobes. It is in this region of the brain that humans process information and make decisions. All our plans are cooked up here.

Besides this intriguing brain development, *Homo rudolfensis* also had a flatter face and a much larger body. The male of this species stood about 5 feet 10 inches tall which, along with his flatter face, would allow him to fit in, or at least not look too threatening, in a crowd at Trump Tower. Also, it is possible that his etiquette would be more conventional while rambling through Central Park. No climbing trees for fun: *Homo rudolfensis* was a full-time ground dweller.

There are other features which seem to be missing, from neck to toe, in the fossil skeleton of *Homo rudolfensis* which make this species somewhat controversial as far as categorization. If the paleoarcheologists knew for sure that "the knee bones' connected to the thigh bone," researchers would be more united in their conclusion about the classification in which to place *Homo rudolfensis*. For modern humans, the warning signs are there: if you would like your fossilized skeleton to make a spectacular splash as a relic, think about a position at death that will keep all those bones in place. One out-of-place bone might relegate you to an anthropological question mark.

Homo Ergaster/Separate or Not Separate

Homo ergaster lived 1.6 to 1.5 million years ago and was also a neighbor to *Homo habilis*. Since stone tools were found nearby fossil skeletons of *Homo ergaster*, his nickname is "Working Man". At that time in eastern and south Africa, *Homo ergaster* was a group who were just trying to make a living among other types of the genus *Homo*. It is not agreed upon as to whether *Homo ergaster* is an entirely different species; the dates of Working Man not only overlap Handy Man but also the subject species of our next inquiry...*Homo erectus*. It is because *Homo ergaster* could not find a better time to enter the scene that has caused many modern researchers to question where this species belongs in the lineage chart of humans.

Turkana Boy

The most complete early fossil skeleton of the human family is called Turkana Boy. Scientists know many things about this skeleton. They know he was a boy of about 8 or 9 years old at point of death. He stood 5'3" and weighed 106 pounds. His fossilized skeleton was found lying face down, among the reeds on the floodplain of the ancestral Omo River in Kenya. This river once fed the nearby modern Lake Turkana in east Africa. Turkana Boy's skeleton bones, though almost 50% complete, were still somewhat scattered. Amazingly, there were no marks of gnawing by canine predators on his skeletal frame. His skull, which was found upside down embedded in the base of an acacia tree, had served for millennia as a subterranean flowerpot.

Turkana Boy lived approximately 1.6 million years ago. The cause of Turkana Boy's death will never be known. He had lesions alongside two of his teeth in his lower jaw. Some researchers think the lesions might have caused abscesses that in turn caused fatal blood poisoning. Other scientists point to an abnormality in his spine called a lumbar disc herniation and think this might have been a possible cause of his death. Either way, the evidence of such disease, abnormalities, and structural problems in a very young person means that before the time of his early death, the child probably suffered a great deal. There is a lot to be said for modern medicine.

From Shoulder to Leg/Strikingly Modern

Below the neck Turkana Boy is essentially human in form. He resembled a preadolescent male of our own species with a tall body-shape well adapted to dry conditions. He had remarkably long legs. His long appendages served him well for someone who was living so close to the equator. All animals must find a way to shed heat. We humans shed heat through evaporation by sweating. The longer the arms and legs, the more the surface area there is for cooling. Because his legs are much longer than many modern humans, it is presumed that he walked and ran great distances in his everyday life.

Above the neck Turkana Boy's skull tells a somewhat different story. His skull and jaw look like an archaic human. If Turkana Boy were alive today, his face and head would alert modern people that something was amiss. He had a sloping forehead and flattened braincase with a heavy brow ridge. He also had a low ridge at the crest of his skull which served as an attachment for large facial muscles. His face was wide and short. In every way, his facial traits are very similar to the genus of *Homo* who came before him. Turkana Boy, so tall for his age, with such long, strong legs, looked like a human with a touch of *Australopithecus*.

From *Homo ergaster* to *Homo erectus*

Turkana Boy is part of the *Homo ergaster* lineage. This species is variously thought to be ancestral to, or as sharing a common ancestor with, or as being the same species as, *Homo erectus* (*Homo erectus* is the subject of our next chapter). The issues between researchers come down to taxonomy. To date scientists can't agree whether *Homo ergaster* is a *Homo erectus* or a separate lineage. Other researchers think that *Homo ergaster* was the predecessor to *Homo erectus* and was therefore the species that spawned the later humans.

Presumably, *Homo ergaster* knew what lineage he came from. Despite the kerfuffle now, he probably wasn't "his own grandpa". What is true to say is that some very ancient humans lived side by side in East Africa for several thousand years. What these various hominins thought about each other is a mystery. Their connection can be identified, but their judgment of each other cannot. Possibly, as the thousands of years went by, they got used to each other. Since our lineage is not nearly that old, we haven't a way to pass judgment. Though in looking at warring tribes among our own species, it doesn't bode well for anything that is directly related to us.

A New Creation Story

As was mentioned earlier, with each new hominin fossil find, it appears that the earlier accounts first posited by Darwin about the origin of humans have been too simple. Researchers have questioned the models for the human family tree—from a single trunk, straight as a Ponderosa pine, up from *Homo habilis* to *Homo sapiens*. Also, dates of our origin ancestors are being pushed back faster than anthropology textbooks can keep up with to print the new data.

These discoveries do not challenge the ultimate baseline of Darwin's theory, but its tidiness. The thinking is that our origin should be looked at as a multi-branched bush, rather than a straight pole of a tree. When looked at as a bush, it seems clear that not all hominin species' DNA managed to reach our own ancestors: some species died out without much of a trace and took their DNA with them.

The Four Fs

The simpler story is that the Great Rift Valley of Africa is where modern humanity arose. Originally the spotlight shone on human's African origins starting with *Homo habilis.* The non-controversial part of the story is that *Homo habilis* was molded by the harsh and threatening environment and who, therefore, was shaped by the "four Fs" of evolution: fighting, fleeing, feeding and mating. This evolutionary progression is not questioned. But the idea of the distribution of human ancestors as all coming from a single common antecedent has been investigated.

Mitochondrial Eve and Chromosomal Adam

Scientists believe that modern humans come from a small set of East African ancestors and everyone is part of the same family, sharing a mother called Mitochondrial Eve. When a sperm fertilizes an egg, only the mom's mitochondria survive. A mom's mitochondria are passed down unchanged and can be traced back from daughter to mother ad infinitum. Mitochondrial Eve living between 230,000 to 100,000 years ago, is as famous in the scientific community as the Eve of the Old Testament is in other communities. Mitochondrial Eve, one woman, is the mother of all humans! And she didn't put up with any ribbing.

The Y Chromosome is also unique. Only males get the Y-Chromosome. Sorry ladies, but daughters are not included. And, like mitochondria, the Y-Chromosome is passed on from father to son virtually unchanged from generation to generation. Chromosome Adam lived between 350,000 to

200,000 years ago. Though the ancient granddad ancestor is called" Adam", scientists admit that there is no way to know if DNA "Adam" and "Eve "ever met! This puts a new spin on the Garden of Eden.

As of this printing, most researchers find that the conclusion of an African lineage is essentially correct. Newest research backs up the idea that our archaic human ancestry evolved around 350,000 to 200,000 years ago (latest findings might someday push back those dates) in Africa and that all living people are related to this small group of humans.

Considering the amazing new scientific developments and the fact that all over the world researchers are unearthing new findings, nothing in paleoarcheology and paleoanthropology is a slam-dunk certainly.

Why Did We Need to Be Smarter?

After our ancestors swung down from the trees, they remained upright on two legs, but though they existed for about 4 million years, all *Australopithecines* have long ago vanished. During that time, some were changing, and among other things, developing bigger brains. The size of hominin brains constitutes the major development in the rise of humans.

It took millions of years of evolutionary change, but about 200,000 years ago hominin features were distinctively modern; the brain had moved forward, and the face moved under it. The forehead appeared, and the heavy brow ridge disappeared. The back of the skull was rounded instead of angled and flat. All these changes are related to the development of the brain.

But why, starting from the beginning of human history, did the brain of hominins continue to increase in size? Why, in other words, did we need to be smarter?

Some of the earliest hominins, probably *Homo habilis*, who had been principally vegetarians and opportunistic scavengers, got the urge for more meat in their diet. More meat equals more calories. To become successful at hunting, rather than primarily gathering, hominins had to band together cooperatively. One human in the vast wilderness with a sharpened stick can't become a robust, successful hunter of big game. Yet many humans grouped together can change the course of history. And so, they did.

A Division of Labor/Necessary for Survival

The first step in evolutionarily changing from a timid, solitary foraging species into a powerful hunter species-to-be-reckoned-with is to group together and form protected shelters. Off-spring need to have a safe place to grow to maturity and the skill of looking after the young was different from the skill of hunting for large game. Both skills were indispensable for the survival of the species. There needed to be a division of labor. Since necessary skills of individuals within the group needed to be considered, hominins had to start challenging their brains to make decisions which would have future consequences. Thinking was about to begin big-time.

There had to be a sharing of the campsite with individuals who were going to cooperate in hunting, and other individuals in the group who were going to look after and protect the young. A special value started to be placed on personal relationships. It was necessary to know who would cooperate, who might be overly competitive, and who was going to be loyal.

All this thinking began to have its effects. Members of the group who could successfully figure out in advance what might happen were the individuals who were most likely to have a successful mating partner and whose genes would eventually become dominant. After all, these were the individuals who also could figure out which person made the best mate...one who would be honest, helpful and not filled with perfidy. No one back then or now wants to sing Shania Twain's country western song "Whose Bed Have Your Boots Been Under?"

A Weighty Matter: Bigger Brains Equal Bigger Heads

As the brain sizes of the hominins became larger, so did the skulls of the infants. There are several ramifications of having a baby with a large skull; one is that, as an accommodation, the female pelvis became larger than other primates of the same stature. Another consequence is that a larger brain case meant that the baby must come out early; some would say a human baby comes out "prematurely" because it is so helpless and needs so much time for development.

For comparison, a baby giraffe comes out of the mother fully formed after 15 months in the womb. Within twenty minutes after birth the baby, weighing 200 pounds and standing six feet tall, is ready to embark on giraffe life. Other than a gulp of milk now and then, the newborn bravely takes its place and walks with the herd.

New Flash: It Is Usually Painful to Give Birth to a Human Baby

If you are human and reading this, I am sure I don't have to tell you that human children need much more protection and learning time than a baby giraffe before they reach maturity. Human children stay with their parents a long time for shelter, security, and to acquire necessary knowledge. No other land mammal requires such a long period of time to develop before they can even walk. Give or take a few months, the human baby requires fifteen months to begin to walk without parental support. All in all, it seems like the giraffe mother got the better deal.

Though human females have larger pelvises than their more primitive ape cousins, they still have narrow pelvic canals. This is because the structural advantage of walking on two legs will not allow for an exceptionally wide expanse between the two upright supporting legs; legs that are a necessity for locomotion. Women of today with narrow pelvic canals and big hips can blame it on a mechanical design glitch; faint consolation.

The upshot of this is that the shortcomings of a baby with a big skull to hold a big brain means that human females forevermore frequently have painful childbirths. Also, women were eternally equipped with large hips (on average a female's pelvis is wider than a man's) and a child whose length of time at home was going to be vastly increased. We women want to thank you Mrs. Early Hominin. The human babies may be cute as they helplessly coo and smile with their baby arms outstretched, but there is a price to pay for everything.

The Cost of a Bigger Brain/Cognitive and Behavioral Development

The brains of humans have rapidly increased in size in the past 2 million years. In fact, it is brain size that separates humans from other apes in the behavioral department. For instance, humans, though related to chimpanzees, rarely rely on breaking off a stick and shoving it into a termite mound as their major source of protein. Though hominin brains started to increase in size as intelligent thinking increased, there was something in the brain's internal structure, besides the size, which added to human cognition.

There really isn't any way to argue that larger brain size alone equals a smarter human. If brain size were the only determinant, the sperm whales would be building the skyscrapers and driving and flying around in machines. Weighing in at 18 pounds, the sperm whale's brain is six times the size of a human's brain (the adult human brain is about 3 pounds). Even the bottlenose dolphin's brain, at 3.5 pounds, is larger than a human's.

The internal structure of the larger brained humans then, is what caused much of the behavioral changes. As our ancient ancestors went looking for food, they started working to change the world around them by banding together to suit their purposes. Larger brains with new structural changes were causing them to look at a stone for what it could do for them rather than an inanimate object to be trod upon.

With the insight that created behavioral changes, it is thought that Handy Man fashioned implements from more than stone. Though tools made from bone, wood, antlers and other organic materials have not survived the several millennia that followed, there are many paleoarcheologists who argue that these materials would also have been used by more advanced hominins.

The individuals, who lived together within the intimacy that would be a necessity in a small tribe/band, started making selections based on their knowledge of where to find new resources. They were making selections too, about who might cooperate and who might compete. This new kind of thinking; thought processes that were based on social behavior, fed into the group's ability to achieve and ultimately thrive. Those groups whose leaders were the most insightful thinkers also had the most successful outcomes.

If the Brains Say They've Got the Munchies, Better Go Out and Look for Food

However, unbeknownst to them, these clever behavioral adaptations were costing our origin ancestors. There was a charge for the new tissues in the brain and the new neural connections that moved around information. The price was the great amount of energy that had to be diverted to the grey matter inside their head. After all, who's the brains of the operation? To grow and thrive, human brains need caloric energy, energy that helped our ancestors figure out how to fit into and control their environment.

Although the brain was about 2 percent of our archaic ancestor's weight, it consumed 20% of their oxygen and got 20% of the blood flow. These percentages haven't changed and are the same for modern humans. Just like an automobile needs gas to run, a human brain needs calories. The caloric energy that is needed for human brains is unwavering; we need calories (mostly in the form of protein) not just part of the day, but twenty-four/seven. Eating more caloric dense, high quality foods was the price to pay for the privilege of a whirring, prodigious new brain.

Since there was a great need for units of energy to keep an increasingly demanding brain alive and functioning, the search for food became the overriding activity for *Homo habilis* and later humans. Paleolithic food sources were most probably plant based and supplemented with meat. Nuts and seeds, fruits from various trees, legumes, leaves, flowers, and insects were all consumed to feed their expanding brains. This task eventually became too much for *Homo habilis*. Around 1.4 million years ago, the species went extinct.

But there was one species of hominin who figured out how to obtain more quality food fuel in a fairly simple way. *Homo erectus* will be the subject of our next chapter.

(1) *The earliest indirect evidence found of hominins using stone tools are fossilized animal bones with stone tool marks; scrapings that date back two and a half million years.*

Humans and Close Ancestors

Years Ago
700,000 500,000 300,000 100,000

Homo sapiens

Neanderthals

Homo heidelburgensis

Homo erectus

"Homo" is the name for the genus—or group of species—which includes all humans, modern and ancient. One million years ago there was a further increase in brain size with Homo erectus. Homo sapiens evolved over 200,000 years ago in East Africa. Neanderthals went extinct about 30,000 years ago.

Chapter 5: The Most Durable Human Species Ever/*Homo erectus*

The most recognizable human yet to be discussed is *Homo erectus* or "Upright man". This ancient species lived 1.4 million years ago and grew to a varying range of heights, from shorter to taller, much like modern humans. Some *Homo erectus* stood as much as 5'11" and weighed 150 pounds. For a comparison with earlier hominins such as *Australopithecus*, Lucy from the famous Lucy fossil skeleton stood only 3' 7" tall.

A New Body for a More Modern Human

Nor does *Homo erectus* retain any remaining ape-like features to hands or arms. Unlike *Australopithecus, Homo erectus* hunted and gathered food on the ground, leaving the protective branches of the trees behind. The body proportions of *Homo erectus* were like modern body sizes, with longer legs and shorter arms which no longer were adapted to swinging in and making a nighttime nest in trees. Instead, their dimensions were like present-day humans, though their skulls reflect a similarity with "Caesar", the intelligent ape in the movie, *Rise of the Planet of the Apes*. Their skull was relatively elongated, but low, with a flat forehead. The forehead was also broad across its base but narrow higher up. There was a strong bar of bone above the eye sockets. Although the braincase was voluminous, eventually reaching 75% the capacity of modern humans, the walls of the skull were thick and reinforced. The teeth were somewhat large, though human in shape, and set in a thick jaw bone that lacked a chin. Even the bones of the rest of the skeleton were thick, suggesting that life put heavy demands on this ancient ancestor.

The Continentals of Their Time: *Homo erectus*

Though many of the *Homo erectus* who lived in Africa were people who began to journey from distant place to distant place, there were always some non-wandering *Homo erectus*. For whatever reasons, some members of (what is otherwise a migratory) group, don't like to travel and decide to remain behind. Even some migrating birds decide not to follow the flock.

Yet, many *Homo erectus* travelled long and arduous distances.

<u>*Homo erectus* Fossils Found Many Places in the World</u>

The most famous Upright Man skeleton was the one found in Java in the early 1890s. Most experts believe that *Homo erectus* originated not in Java but came from Africa well over a million years ago. Some of the Javanese skeletons are about a million years old, while the youngest are less than 100,000 years old. Chinese *Homo erectus* fossils span a similar date ranging from 1 million years old to 250,000 years old. Over this long-time span in Java, China, Africa and a few other places in the world, Upright Man did some changing of brain size and body size. After their exceptional growth spurt, they then made no more significant evolutionary changes. If they invented some great sauce recipes or drew some cool tattoos on themselves, their secrets died with them.

<u>The Brains Got Bigger During the Time of *Homo Erectus*</u>

With the arrival of *Homo erectus* (1.4 million years to 143,000 years ago) there is a shift in brain size. From 3 million years ago to the present day, brain volumes in hominins in general have increased three times in size from our earliest ancestors. Most of the additional brain growth came during the time of *Homo erectus* starting around 1.3 million years ago.

The very earliest *Homo erectus* had a brain size not too much bigger than *Australopithecus*, but as time went on, the species of *Homo erectus* developed a brain volume that was over 50 percent larger than *Australopithecus*. Yet, compared to modern humans, *Homo erectus* had a brain dimension that was slightly smaller.

Nevertheless, Upright Man had the largest brain available for the new model of hominins. And though *Homo erectus* didn't brag about it, they were the smartest ones walking the planet at that time. With their large body and calorie energy-craving brain, *Homo erectus* needed to constantly bulk up on high quality, high density food.

The Greatest of the Early Inventions: Controlling Fire

It is thought by many researchers that it was *Homo erectus* who first discovered how to use and control fire. (1) The controlled use of fire was, of necessity, one of the earliest of human discoveries. And it might be the most important breakthrough ever. The rumor that *Homo erectus* figured out barbeque is an exaggeration, but cooking is a defining feature of humanity. We moderns haven't found a substitute for our ancient ancestor's hot discovery.

Cooking food has many benefits, but for the bodies and brains of the earlier hominin species, heating and cooking plants and animals gave them a biological advantage. There were anatomical changes in brains and body to respond to: cooked food makes it easy to eat large volumes of tough, fibrous plants by softening them and making them easy to chew. Cooked meat is more digestible for the human gut. And cooking creates meat that is safe to eat by killing harmful bacteria and parasites. The digestible factor went up significantly; all that was needed was to learn to control fire and find an ancient cookbook with recipes that had a short prep time.

Which Came First, The Brains or The Cooks

The management of fire required an ability to conceptualize the idea of the controlled use of fire. This meant thinking about exercising power and responsibility over an outside force. A larger brain with more cognitive functions would have been needed. Yet, it is impossible to know precisely in what way our ancient ancestors more than a million years ago discovered how to control fire. It is a chicken and egg problem: cooked foods were needed for bigger brains, and bigger brains were needed to figure out how to cook foods.

Adapting to Cooked Foods

Homo erectus was originally a foraging species. They gathered more than they hunted. In fact, all we humans descend from scavenging populations.

By the time Upright Man was walking the planet, even their mouths were beginning to be shaped in a way that looked more like members of our own species. *Homo erectus* had teeth that looked strikingly like ours. Like modern humans, their jaws were not strong compared to other great apes; the muscles connecting their mandibles were weak. Researchers point to weak jaw muscles as evidence that our ancient forebears had adapted to cooked food. Not only would tough foods become soft and easy to chew, but most importantly, more food could be consumed.

Just how fire was first discovered is up for debate. The best guess is that *Homo erectus* had an opportunity to see how fire cooked food by observing spontaneous occurrences (lightning strikes, meteor impacts, forest and grass fires caused from drought, sun, etc.) that naturally heated up the vegetables they foraged.

Mashed, Fried or Boiled?

Once the fire had passed through an area, Upright Man went to the location where his group always dug for roots and tubers. Ancient roots and tubers were a popular item on the menu; though potatoes did not exist at that time in Africa, the ancient roots like turnips might well have tasted like potatoes. After a fire, tubers were dug up and the root that had been difficult to chew when raw was transformed. Your guess is as good as mine as to the flavor, but it was sure to taste better cooked than raw.

Who could resist the idea that cooked was better than uncooked? Even today, all over the world cooked food is preferred over raw food. There is no culture that specifically relies on uncooked food as their staple diet. A preference for cooked foods has also been observed in chimpanzees and great apes. Make that a burger with fries, hold the catsup.

Pass the Pots, or Maybe Just the Stones

Exactly how early humans were utilizing fire to cook their foods is not known. Most researchers believe that they were simply laying food on top of a large outdoor fire pit and thereby creating what we call roasting.

Other researchers think that early humans may have boiled their food. Ancient stones have been found that show evidence of being heated. After the stones are hot, they can be transferred to a primitive cooking container made from hides, gut, ostrich eggshells or even the stump of a dead tree trunk. The water and other contents inside the container then comes to a boil with the aid of the heated stones. Maybe the first chicken soup was invented in the shell of an ostrich egg with hot stones heating up the broth. Oy!

Food Processing the Old-Fashioned Way

There are other primitive methods besides cooking with fire that process food and make it more edible for human digestion. Some of the methods that might have augmented foods for cooking could have been winnowing, cutting, pounding, peeling and/or fermenting. There may have been master Stone Age chefs who specialized in fixing indigestible plant parts and tough meat, changing it into flavorful concoctions, and in the process making digestion easier. Perhaps the foodie movement started earlier than we think. Pass the salt.

Other Warm Advantages to Fire

There are many other advantages to using fire: it adds light and heat to the night, fire clears forests, it heat-treats stone tools, and fire keeps predator animals away. For pulling together a group, fire also has social purposes: gathering places can be welcoming with a central fire. If some members are away from their group, a fire can be used as a beacon. Since everyone has their own distinctive sleep pattern, there were probably always a few night owls in the group who would want to stay up after dark. Everyone would have a choice as to what the bedtime hour would be. And finally, it is thought that *Homo erectus* began staying seasonally or even semi-permanently in cool places. There is nothing like chestnuts roasting on an open fire when your group has decided to make a long stay-over in Frosty-Ville.

How Physical Changes Can Occur When Fire Is Available

In the long line of anatomical changes that were taking place in early hominins, some were directly related to the eating of cooked, soft food. *Homo erectus* evolved smaller mouths, smaller teeth and smaller digestive systems than either *Australopithecines* or *Homo habilis*. In fact, the chewing teeth of Upright Man are small compared to all other primates. Researchers theorize that this physical change might be due to the cooking of food for over one million years: softening it enough so that strong jaws and large teeth for ripping, tearing and chewing would not be necessary.

Not just teeth, but stomachs in the *Homo* genus also became smaller in size; 97 percent smaller than other primates by body size. The same thing is true for intestines and bowels. It is thought that humans evolved to have small guts as an adaptation to eating cooked foods. With the advent of cooking, indigestible fiber was softened enough so that our comparatively small intestines were not overworked. As our large intestine (colon) developed, it became much smaller than other primates. Though our colon cannot retain as much fiber as the great apes, if most of the food is cooked, there isn't the need.

Large amounts of raw meat are impossible to eat unless they are first cooked. There are two reasons for this: humans don't have the mandibular jaw power to eat massive amounts of raw meat all at once, and the human digestive system can't handle excessively large portions of raw meat.

Fat: The Gift That Keeps on Giving

The benefits of having a smaller gut couldn't be realized until high-quality, high caloric dense foods were available and present all year long. The necessary steps for our cooking ancestors were to stay in groups and hunt and forage cooperatively, and to control fire and use fire pits and rudimentary containers effectively.

Another benefit of cooking that is directly related to roasted meat is that cooking meat also makes it easier to store animal fat as body insulation. In studies of people eating a high fat meat diet, the participants could achieve the same weight gain as plant eaters who were eating five times the number of calories in the form of carbohydrates. Back in the earliest historical period of human history, when fat was such a valued commodity, there probably would have been very few who would have joined a Stone Age Weight Watchers Club.

As most of us already know, fat cozies-up in layers around the middle part of the body where the digestive system is located. In the digestive system, fat cells absorb some important nutrients. The story is that one of fat's greatest benefits is that it gives off energy. Also, fat acts like a little furnace, protecting organs by warming and insulating them from the cold. But couldn't wearing a nice fluffy fur bear skin do the same thing?

Meat Can Be Toxic If It Is the Primary Food Source

The need for protein to stoke the larger, energy-hungry brains doesn't mean that once fire was controlled, there was a switch to a disproportionately high meat diet. *Homo erectus* was not the hunter that later humans would become. And there was not a need for excessive animal meat consumption. Humans need carbohydrates (from plant food) and fats (found in nuts, seeds, and animal flesh). Without carbohydrates and fats, unbalanced protein consumption will induce a stream of nitrogenous waste material to form, leading to dire consequences. Digesting excessive meat protein induces a type of blood poisoning with symptoms that include toxic levels of ammonia, dehydration, damage to kidneys and liver, and risk of possible death. However, since *Homo erectus* is the most successfully long-lived of the genus *Homo*, it is likely that they did not exceed their maximum safe level of protein which is around 50 percent.

The Healthy Meat of Ancient Humans

You may or may not have read the *Eat-a-Bug Cookbook*, (available through Amazon and your local bookstore), but ancient humans could have added their own recipes to this recently published hardback. In the cookbook there is a very tasty recipe for braised tarantulas. The directions are, first you need to catch some. Then, after freezing the tarantulas, removing their large stomachs, and burning off their hair, the spiders are dipped in a tempura sauce and deep fried.

Obviously, this modern technique would be beyond the technical capacity of Stone Age humans. But that is not to say they would have balked at the idea of eating spiders, bugs and insects. Caterpillars, worms and grasshoppers were undoubtedly eaten. These are reliable sources of protein and much easier to catch than large mammals. Insects could be called the lazy hunter's white meat.

A Recipe Made for a Hungry, Big-brained Human

Here is a possible recipe from the Stone Age. Ancient peoples might have said it is better than potato chips:
First, find a salty, ancient lake;
Second, find young grasshoppers whose wings have not yet formed;
Third, with sticks, drive grasshoppers into the salty lake;
Forth, gather up the drowned grasshoppers;
Fifth, place grasshoppers on flat stones to sun-dry;
Sixth, after they have dried…EAT.
Bon appetit!

Hair No More or Naked by Design

Mammals are warm blooded, but not so much so that they enjoy going without their fur. Ask your dog or cat if they really want to part with their furry attire. Perhaps they will give vague animal answers, but when trying to shave off every scintilla of their fur, expect them to launch complaints.

Some mammals have little or no fur, but they have skin variants that compensate for their loss of hair. Rhinoceroses and elephants, for instance, have crusty, thick skin that allows their bodies to retain heat at night and cools them off in the daytime.

Skin is the largest organ of the human body. When an adult's skin is weighed, it averages 20 square feet and weighs almost 10 pounds. Humans, with their thin, delicate skin, have somehow been reduced to wearing a large but fragile and easily penetrated epidermis. In fact, humans are the only hairless primates on the planet. How and why did that happen? Theories abound. Since scientific conclusions differ, you will have to choose which one or maybe several of the theories are most appealing.

Hairless Theory Number One

Why we are hairless: The water theory suggests that 6 to 8 million years ago our ancestors, (some creatures like *Australopithecines*), did most of their searching for food in the East African shallow rivers and lakes. We lost our hair because fur is not an effective way to stay warm in water. Instead, we stored body fat as our major heat source. Since body fat doesn't fossilize, we can only use our imaginations to picture naked, archaic hominins splashing away, catching and eating fish to store as fat and use as fuel for their growing brains. Of course, if you would prefer not to use your imagination in this case, c'est la vie. This theory doesn't explain how we stayed warm at night, but it does explain how we have slight webs between our fingers.

Hairless Theory Number Two

The second explanation for humans' lack of fur is the down-from-the-trees theory. As the ancient hominins were leaving the trees in the forests in the daytime to forage in the hot African savannah, they needed to ditch the hair. Too much fur on these bipedal hominins as they hiked through the open savannah would cause overheating to occur.

Hairless Theory Number Three

This theory is tied to theory number two. It agrees that hominins lost their fur because they went out in the hot African sun to forage and, in later times, to hunt. This is the perspiration hypothesis. This theory starts with *Homo erectus*. *Homo erectus* had a body that developed an athletic frame; a frame that was adapted to both hunt prey and avoid predators. Their athletic body included long Achilles tendons, narrow waists, broad shoulders with arms that swiveled for throwing, and a head that rotated independently (to watch for prey and predators). *Homo erectus* was built for running. Extreme running on the open savannah created an Upright Man who went hairless. And, a running hunter needed to sweat.

Sweating dissipates heat through evaporation. In this theory, we humans lost our heavy coat of body hair and gained more sweat glands so that while running, either toward prey or away from predators, the sweat glands would keep our bodies cool. Losing fur and sweating helped humans to "dump the heat": especially heat that would build up in their larger brains which, when they overheat, tend to cook inside the skull.

When searching for food on the hot, dry savannah, the humans with more sweat glands could stay out longer and find more edible nourishment before retreating to the shade. The longer the humans with abundant sweat glands and no fur could hunt and forage for their families, the healthier their offspring would be. Evolutionary pressures selected for less hair and the ability to perspire.

Not that everything went easily for early humans. In the beginning, the sweat glands and abundant fur on the archaic humans were located, among other places, on the inner surface of their hands and feet. It was a slow process to rid dad and mom of their furry, sweaty palms and soles. As evolutionary processes continued, selection of traits that enhanced the survival of this species were determining for less hair and sweat glands on hands and feet. Through successive generations, palms and soles became smooth and less sweaty, and more perspiration glands appeared in other body locations. By the time evolutionary factors had played their part, humans would have shiny, hairless soles and palms and more than 2 million sweat glands in other locations in their body. This is the same number as modern humans. If you don't believe that, just count your sweat glands.

While other animals sweat, humans are the world class, top-notch sweaters. Among all primates, humans have the highest-density sweat glands. To our credit or our embarrassment, humans are not only the hairless apes, but the sweatiest primates on earth.

Hairless Theory Number Four

This is the "ridding humans of the little beasties theory", also known as eliminating body parasites. I have written elsewhere of the need for humans to cook meat to kill parasites. But there were also external parasites that evolved as humans were evolving. These external parasites found a cozy home in our hair. Human lice infections are confined to the hairy areas of our bodies. Researchers have done DNA dating on ancient parasites and there is a correlation between the evolution of the human body louse and the amount of human body fur.

We are talking here about human parasites; blood-sucking lice, ticks and fleas. Who wouldn't want to cut down on these free-loading creatures? Our early ancestors may have lived in ancient times, but they knew a parasite when one started rapaciously drawing out their plasma.

Homo erectus built fires, constructed shelters, used animal skins as cloaks, and generally cared for themselves in cold weather. Losing their body hair was not too significant. And the benefit of no fur would be the reward of reducing the number of body parasites. The relief would have been palpable. So, humans, far from being creatures who felt regret at their nakedness, (as we might infer from Biblical stories), might have been content that evolutionary expediency influenced the no-fur determination.

Some Problems with the Eliminating-Fur-to-Rid-the-Body-of-Parasites Theory

This theory doesn't explain what role the retention of hair on a human head might play in hominin development. Nor does it clarify what the significance of baldness is in the long history of human evolution. The theory is justified this way: hair on the head is a protection from the heat of the day and the cold of the night. But that same reasoning could apply to hair on the body; which would lead us in a circular direction, back to the question of why we lost our fur.

Perhaps the retention of head hair has to do with the attractiveness idea: head lice be damned, we still want to look good.

The Real Existence of Pubic Lice or the Imaginary Existence of Sexual Odor

The theory of purging the body of lice and ticks has an answer for the continued existence of pubic hair. Pubic hair was retained as an odor enhancer: other mammals give off airborne pheromone scents for sexual attraction, so why not humans? Pubic hair would be the odor dispenser. The only thing wrong with the pubic hair theory is that there is no evidence that human pheromones, exist---but unfortunately specialized pubic lice do.

Selecting for Sex Appeal and Smooth Skin

As nakedness let everything show, smooth, hairless skin might have been a magnet for sexual attraction. Females were not the only ones selecting for mates, males were too. From archaic human females up to modern times, on average females have less hair than their male counterparts. Some researchers theorize that it is possible that for millennia, males preferred and selected females with less body hair and that this is now a significant determining factor in our DNA.

Women: Show Some Skin

It is possible that the deep memory from our ancestors has meant, and still means that hairless skin is more appealing in women because it signals that they are free of body lice and ticks. Certainly, deep cut evening dresses and skimpy party frocks expose a lot of female flesh. But the reverse, that is skimpy apparel for men, doesn't seem to be the case. So, either the theory of smooth skin's appeal needs an overhaul, or women have always been fond of hairy males who may or may not be carriers of body parasites. The answers are buried in the primordial part of our brains having to do with sexual attraction.

One researcher is so sure that female hairlessness in modern times is an element in our innate sexual selection process that he thinks women shaving the hair under their arms and on their legs is the "last echo of an ancient instinct". He may or may not be correct, but investing in shares of hair removal creams, mousses, gels and waxes is still a safe bet on the stock market.

The Dark and the Light of Skin

Humans develop skin color much like other mammals. Special skin cells are packed with pigment molecules. The more pigment, the darker the skin. The genes for light and dark skin color are over a million years old. It is speculated that all hominins were originally light skinned when they had their fur. This is the case with our distant cousins, the chimpanzees. Their skin is light underneath their hairy coat. But since they are fully covered, there is no need to have protective, dark skin.

As soon as hominins started losing their body hair, they would have needed dark skin for protection against the sun. On the hot savannah, humans who had a gene for dark skin would have had more chances for survival. With the loss of hair, within a few generations the skin of *Homo erectus* would have started to become darker. This darker pigmintation and loss of body hair probably occurred from one to two million years ago. At that time we lost our hair and gained an evenly dark pigmentation.

People who lived with intense ultraviolet light benfited from dark color pigments that shielded their skin. In places with less sunlight, people needed lighter skin to absorb more sunlight.

Over 1.7 million years ago, hominins had left the protection of the tropical forests and started living on the open African plains. At about 1.2 million years ago, it is estimated that the hominin population consisted of only 14,000 breeding individuals. At this time, researchers have suggested that all hominins had lost their hair, gained the necessary sweat glands, and had dark skin.

From an evolutionary standpoint, the important factor is to protect the skin's DNA. Since hominins first developed in the hotter latitudes, the protection factor had been in full force for a million years by the time our modern species (*Homo sapiens*) came along. The theory is that our species, which developed over 200,000 years ago, were originally all dark-skinned.

Skin Cancer and The Color of Skin

There are several popular theories as to why humans developed dark skin. Since these various theories are not necessarily mutually exclusive, it is possible that more than one explanation is correct. The possibility of contracting skin cancer is one theory that has been advanced.

Exposure to too much direct sunlight can cause skin cancer. Earlier hominins shed their shaggy hair, exposing their light skin and at the same time, started to develop a form of skin cancer. On the sun-drenched savannah of Africa, natural selection favored those whose skins were darkening for protection against ultraviolet radiation (UVR). Those who had the darkest variations in skin color not only had protection against skin cancer, but also had the healthiest offspring whose skin could also ward off cancer. Again, the natural selection process would choose those offspring who would have the greatest genetic advantage to pass on their genes and allow their species to thrive.

Though this theory is disputed by some researchers, other scientists believe that skin cancer prevention was a driving force in human evolution, permanently darkening the light-skinned humans.

Ochre: Body Painting or Protection?

Ochre use is among the oldest coloring that is known that humans purposely used. Ochre is an iron ore that is red in color. Its use goes back to the earliest Paleolithic period. About 400,000 years ago, *Homo erectus* seems to have been chipping off ochre from local rocks in South African caves.

A hundred thousand years later, at a site in Terra Amata in France, several pieces of ochre were found with primitive stone tools. The lumps of ochre showed unmistakable signs of wear marks. This has lead researchers to believe that *Homo erectus* had used these pieces of ochre; though what they used them for is unknown.

Ochre is not only red, but frequently its color intensity is like blood. Because of the resemblance with blood, there are theories that ochre might have represented some symbolic relationship with gore or body fluids. The practice of body painting is a very ancient tradition and though it can only be speculated, it is possible that, in lieu of hair, the canvas of the human body became too tantalizing to resist.

There are people in remote parts of Africa who even today mix ochre with grease, slathering their bodies to protect them from sunburn and insect bites while at the same time giving themselves a cosmetic body make-over.

Homo erectus: Dmanisi

Though it is believed that *Homo erectus* originated in Africa, the earliest dated fossils of *Homo erectus* are not found in Africa. In recent years, fossils of *Homo erectus* have been found in Dmanisi, in the Republic of Georgia. These fossils have been found to be

1.4 million years old. The Dmanisi hominin fossils are the oldest hominin fossils so far to have been found outside Africa.

The skull found in Dmanisi was of an older person. The fossil shows that the person had lost most of their teeth many years before death. The jaw bone had deteriorated to the point that the person would not have been able to eat without help from others. Researchers believe that members of his tribe kept him alive by softening or liquefying his food. This, researchers think, might be an illustration of one of the first examples of acts of human compassion.

A Traveling Man Out of Africa

Enough research has been done at Olduvai Gorge in Tanzania to lead paleoanthropologists to believe that somewhere in that area of East Africa is the seat of human origin. At Olduvai Gorge there are primate fossils that go back almost 25 million years. But if that is the case, what is a very early *Homo erectus* fossil doing in Dmanisi which is essentially in Asia?

Homo erectus was a durable species which survived over a million years. During that time, they must have become bored with continually living in the same place because their fossilized remains have been found as far as China and the island of Java in Indonesia. *Homo erectus* was the first member of the genus *Homo* to make their way out of Africa. Walking northeast and skirting Western Europe, they traveled far to the east. There were populations of *Homo erectus* that went all the way through Southeast Asia to what is now the island of Java. There are many fossils of *Homo erectus* found in Indonesia and in China. You might remember the names, "Java Man" or "Peking Man". That was *Homo erectus* who was creating quite an impression with the discovery of their fossil bones.

The Disappearing Peking Man Fossils

The *Homo erectus* fossils that were uncovered in Indonesia are still available to study, but the fossils from China are not. Although Peking Man fossils had survived for more than a million years, the scientist who discovered them feared hostilities from the Japanese occupation forces and in 1941 he decided to relocate the fossils from Beijing to New York. Sadly, and ironically, the Japanese captured the ship on which the fossils were located, and the Chinese fossils were never seen again.

Some Successes Made by *Homo erectus:* Reed Rafts

Upright man is not credited with greatly advancing his culture or technology. But there were certain things at which *Homo erectus* excelled. As mentioned earlier, this is the longest-lived of the human species. Nor did *Homo erectus* stay ensconced in one safe place but instead, traveled the globe by foot. Many researchers believe that *Homo erectus* also fashioned rafts, though no rafts have survived. It is thought that the flimsy type of material such as reeds and buoyant wood that would have been used to create rafts would have quickly deteriorated.

In each new environment, *Homo erectus* was able to figure out how to cope with their new surroundings. Would our species, under such new, primitive and inhospitable circumstances, manage to find an ecological niche in such mystifying places on earth?

Shells as Tools and As A Possible Art Canvas

Homo erectus is particularly notable for tool creation and usage, but before we get to their ability to wield stone axes, a word should be said about how they used shells. Shells on the island of Java that have been manipulated by *Homo erectus* have been found to date back to half a million years. The shells were found at their encampment and many of the mollusk shells show signs of being pried open for eating purposes. But one of the shells dating to that same period that was found recently shows the undeniable signs of carved abstract markings.

This incised shell design is so far the only evidence that *Homo erectus* may have been interested in creating art. The beautifully carved zigzag markings are the work of someone who wanted to express themselves. What it is telling us, no one will ever know, but it does show that our direct ancestors, 500,000 years ago, were interested in engraving.

The Flintstones and Handaxes of Long Ago

Over a million years ago, *Homo erectus*, our ancient ancestors, were walking the earth and creating stone tools. To make their tools, the types of stones they preferred were flint, chert, obsidian and lava. One or more of these rocks are available in numerous places in the world. Check your backyard.

Homo erectus eventually became outstanding craftsmen who used sturdy hammer stones to reduce flint and other rocks into useable tools. At first *Homo erectus* used large stone cobbles which were roughly hewn. From the beginning, both sides of the rock were crafted, ("bifacially worked"), into a variety of shapes: narrow and thin, flatly oval, or almost circular. They also developed a teardrop-shaped stone that fit snugly in the hand.

These early stone tools eventually were worked into the much more refined handaxes that you might have seen in Hollywood movies dealing with prehistoric people. You could look up a movie starring Raquel Welch called *One Million B.C.* The 1966 movie will never become dated since it is about antediluvian humans. In the movie the buxom and perfectly coiffed Raquel takes up a handaxe for protection. Hollywood often takes liberties with their props, but it is true that the oldest handaxes are very "prehistoric" and date back to over a million years ago. It probably isn't true that females a million years ago looked like Raquel Welch. It was a time when gravity still ruled and cosmetic specialists had yet to figure out how to make surgeon's tiny stone implements.

As the thousands of years went by, later handaxes were much more skillfully constructed and polished with straight cutting edges, reflecting the hours of time-consuming work that went into creating them. Unlike *Homo habilis*, whose more primitive stone tools were used for smashing, beating and butchering the flesh of small animals, or digging up roots or cutting down plants, researchers believe that the handaxes of *Homo erectus* were used in the hunt to dismember and deflesh much larger animals such as mammoths and woolly rhinoceroses.

The tools of *Homo erectus* were used in conjunction with hunting skills. Scientists have found hundreds of flaked tools, including handaxes, lying among the skeletal remains of giant, now-extinct baboons and wooly mammoths. The skeletons of the animals are dismembered, and the meatiest sections of the animals were the parts that were consumed. It would have taken great proficiency by many members of the tribe of *Homo erectus* to bring down such powerful animals.

There is, however, some disagreement among archaeologists as to what the primary function of the handaxe of *Homo erectus* was used for. Were they primarily used for hitting as a weapon for small animals or defensive weapons used against other tribe members? Were they used as a smashing instrument for getting marrow out of bones? Were they used specifically for cutting? Or were they display ornaments used for social and sexual signaling? Any updated news flashes on the latter use will surely be broadcast on all social media venues.

Crushed Skulls Found: Violence and Cannibalism in Early Humans

Though not used to bring down large game, the handaxes used by *Homo erectus* were dangerous and deadly. Many *Homo erectus* skulls fond in India, for instance, are riddled with head wounds. These wounds could be the result of big game attacks, but more likely they were the result of *Homo erectus* warfare with others of their species.

There may have been a dish on the menu that was only served occasionally and not necessarily broiled, roasted or chargrilled. *Homo erectus* himself may have sporadically been a dinner item. A disturbing number of *Homo erectus* skulls dating back 500,000 years have been found in a cave in Longgushan, China. The *Homo erectus* skeletons are nowhere to be seen except for their skulls which have been severed from their bodies, polished and preserved.

Other evidence of possible human flesh-eating are the cut marks on *Homo erectus* bodies in South Africa that show evidence of having been butchered much like other fauna.

Homo erectus: A Puny Meal

There is recent research that questions the idea that cannibalism is for serving up flesh on a dinner plate, or that cannibalism was a wide-spread practice of early humans. The research is based on the size and number of calories each *Homo erectus* would yield. A modern-day American human contains approximately 144,000 calories, but a *Homo erectus* would have less caloric content. Based on the amount of effort that would be necessary to capture and kill enough *Homo erectus* to make a feast for a tribe, the researcher asserts that it would have been much easier to capture and kill a wooly mammoth. When dealing with people who are about the same size as the hunters and having equal intelligence, most *Homo erectus*, the researcher claims, except for tribal rites after a battle, would choose a less formidable creature to hunt and eat.

Though some researchers are suspicious that cannibalism may have been in existence in various *Homo* groups for over one million years, there is no definitive evidence that makes the case that our ancient forefathers ate their own species. Hannibal Lecter, the modern-day equivalent of a cannibal, is quoted as saying, that, "A census taker once tried to test me. I ate his liver with some fava beans and a nice Chianti." If *Homo erectus* didn't hunt and eat his fellow species, perhaps it is because fava beans were few and far between and red wine had yet to be invented.

Cooperative Hunting and Diet Patterns of *Homo erectus*

Researchers have found that the types of animals later *Homo erectus* were eating were frequently not scavenged, but instead hunted. There is a huge butchery site at Olduvai Gorge in Tanzania where wildebeest and other large herd animals were carried, butchered, cooked and eaten. Since the animals were large, it is believed that there was cooperation in the hunt for these creatures. The bones at this site date back over a million years. The meat was stripped from the animal's bones and then cooked. It is probable that the prey animals were hunted in a thought-out, cooperative manner, possibly with the early humans sitting in trees and waiting until the herds were directly underneath. Then, *Homo erectus* at a certain signal would use their wooden spears sharpened to points, to assault the animals all at once and at point-blank range. Other stabbing methods may have included using hand held saber tooth cat fangs. Fangs from these extinct tigers have been found nearby *Homo erectus* fossils.

Crushing the skulls of the hunted prey animals was a popular method for retrieving the highly protein-rich brains. Because of the many crushed skulls of prey animals that have been found near *Homo erectus* feeding sites, it is thought that the brains of animals were one of the delicacies on the bill of fare.

Cooperation between these ancient humans would mean that more meat and bone marrow would be available, which in turn would provide more energy-rich protein to fuel their growing brains and higher energy levels.

Cooking and the Brainy *Homo erectus* Tribes

It takes about an hour for a chimpanzee to absorb 400 calories of raw meat. *Homo erectus* and the later humans could absorb that many calories in just a few minutes if the meat was cooked.

It is possible that through cooperative hunting and the control of fire to cook raw meat, the once solitary gatherers and scavengers were slowly becoming communal and group-oriented. Cooked meat was a great advancement for sustaining their ever-growing vitality. Cooperative hunting helped to make that short eating time possible.

Housing the Old-Fashioned Way

There is no record of the first housing that was constructed so long ago. Possibly, coming down from their aerial nighttime shelters, the earliest hominins used leaves and branches from trees and shrubs for protection. There is evidence that dates back 400,000 years that links *Homo erectus* to the construction of 50-foot diameter huts with rings of stones for foundations that presumably braced tree limbs. It is thought that animal skins covered the frames and possibly the floors. There are remains in Nice, France (Terra Amata) of a site that is the oldest archeological evidence so far of the first shelters and domestication of fire.

Circular bone and stone foundations in East Germany have also been found and dated at having been constructed 350,000 years ago.

Generation After Generation Without Much Change

Although *Homo erectus* traveled far and wide during their long history, they did not advance nor add much to their technology during their lengthy stay on earth (1.4 million years ago to 143,000 years ago). Though their handaxes were very successfully wielded and used in many types of circumstances and environments, for over a million years little changed in their stone tool designs. Their achievements can be summed up as follows: they were very effective travelers, traveling to many continents of the earth; they were able to adapt and survive in new environments; and they successfully improved the stone handaxe. These are the major advances credited to this long-lived species. Their emergence and existence are a period of relative developmental stability or, as some researchers less charitably have put it, their presence was a stage of "evolutionary stagnation".

The species of *Homo erectus* were exceptional in their ability to travel and adapt. Not all of them left their African home. The groups that did leave were probably remarkable and confident hunters, following the massive moving herds. It could be that some *Homo erectus* were more dissatisfied with their location than with the rudimentary form of their technology. What was good enough for their folks was good enough for them. On their bucket list, apparently, "travel to exotic lands" was high in the rankings.

(1) *There is evidence of controlled use of fire 1.4 million years ago in South Africa.*

Chapter 6: The Surprising Species of *Homo heidelbergensis*

Homo heidelbergensis (Heidelberg man) lived 700,000 to 200,000 years ago. Though, like most ancient species connected to early man, there is some controversy as to where this group falls on the human timeline, most scientists believe that Heidelberg man is a direct ancestor to modern humans. The remains of *Homo heidelbergensis* have been found in Africa and are also found in many parts of Europe: Germany, France and Greece are all prime real estate properties claimed by this extinct species.

Though Heidelberg man was prolific in their output of stone tools, most of their axes are developmentally like the stone tools of *Homo erectus*.

Homo heidelbergensis was a robust, strong, tall and well-built species with bodies equivalent to modern humans. Their faces, though differing from individual to individual, were flatter and more human than ape-like. Imagine a rougher, tougher and slightly simian version of ourselves: individuals you wouldn't want to meet in a dark alley.

Brains and Super-Bug Food

The brains of Heidelberg man were almost the size of a modern human's and therefore metabolically expensive. From time to time sparse hunting periods would have occurred. When prey animals were not available they would have had to rely on some type of energy-rich food to fuel their ever-needy neural expansion.

There is a high energy food that fits that description and that has been eaten by humans and proto-humans for millions of years. It is the regurgitated vomit from worker bees which we call honey.

Honey is nectar that has been extracted from flowers and then stored in the stomachs of bees. It has great nutritional properties, just perfect for fueling the energetically needy brains of *Homo heidelbergensis*. Honey is a concentrated source of glucose and fructose and contains trace amounts of vitamins and minerals. Also, honey preserves well and when lathered onto wounds, protects against infections. A taste of honey can be further nutritionally enhanced by eating the honey with the bee larvae (also known as baby bees) for protein, fat and B vitamins.

Bees have existed for one hundred million years. Though there is no direct evidence linking *Homo heidelbergensis* with collecting and eating this nutritious food, wild gorillas and orangutans seek out and consume honey. With Heidelberg man's ever needy energy consuming brain, there is good reason to assume that they would add the special product of the bees to their diets. Honey didn't come in sanitized glass jars. Scooping and licking were acceptable behavior. Though considering the inevitable stings that went with the honey, Heidelberg man probably never got chub-bee.

Quick or Slow Language Theories

Among the changes occurring from switching to an upright position is the occurrence of language. There are many theories about how language arose. There is still a great deal of debate among scientists. Some researchers think that our ancestors started talking immediately as their brains became large.

Ape Gestures

Other scientists think that language evolved slowly from the gestures and sounds used by our ape-like forerunners. In fact, no one knows the chronology of the evolution of language. Unlike the written word, language doesn't leave traces.

Talk Like a Monkey or Talk Like a Human

Though animals communicate, only humans have true language. Since humans are the ones who define what language is, the origin of language is credited to our own species. Chimpanzees, like humans, evolved to have specialized mechanisms for producing sound. They vocalize, but they do not use their tongue to adjust the sounds they make. And because of their physical structure, when apes try to learn to speak like humans, their anatomy prevents their voice box from making anything other than mutterances, hollers and booming noises.

Physical Characteristics Needed for Speech

When did humans start to talk? It is a question that can't be answered. There are no fossil records of speech. Paleoanthropologists have no direct confirmation for the exact time that language emerged. But there are necessary physical aspects to speech built into a human body which makes speech possible. These physical features can be verified.

Heidelberg Man had a physiognomy and throat structure which were consistent with speech. These physical characteristics have not been found in a hominin species before *Homo heidelbergensis*. These changes altered the shape of the mouth, lips, tongue and vocal tract, allowing for a range of coherent sounds.

Say Hi to The Hyoid

Our ability to talk is partially due to several small but significant changes in the human anatomical structure. One is a petite crescent-shaped bone called a hyoid bone perched in the muscles of our neck. The hyoid is shaped like a soft, unsecured wishbone. Other animals have versions of the hyoid bone, but only the human variety is in the right position to work in unison with the larynx and tongue to produce sounds that range from cooing baby talk to interminable political speeches.

Dropping Names with The Help of a Larynx Drop

Alongside the hyoid bone, another important anatomical change happened that, if ancient humans started immediately to use it, could have jolted their sense of what animal sounds are like – it's called the larynx drop.

Human infants need to breathe and drink mother's milk at the same time, so their larynx sits up high in their nasal cavity. It is a trick that you, the reader, cannot perform...try continual sucking milk from a straw (or you could use a more potent liquid than milk) and breathing at the same time. At around three months of age, the larynx in a baby "drops" much lower in the throat, making speech possible, but preventing that wonderful continual drinking and breathing to take place. It's at this time when the child first becomes vaguely aware that at some point in the future they will have to get off the proverbial teat.

There is great significance in the fact that *Homo heidelbergensis* is the first hominin species to have a face and neck that was shaped like a modern human's. It means that Heidelberg Man also had a tongue, lips, and vocal organs that were similar in design to our own.

Hyoid bones and a dropped larynx are some of the foundations of human speech. And these anatomical changes are found in *Homo heidelbergensis*. Based on discoveries such as fossilized hyoids, it is probable that Heidelberg Man had the capability to speak as early as 300,000 years ago.

Waking Up to Talk Like an Animal or The First Animal Who Could Talk

Homo heidelbergensis caught a rare break; all the physical parts necessary to make human sounds were there. And practically speaking (pun intended), as soon as speech became anatomically possible, it would seem reasonable that language would be advantageous. But even if Heidelberg Man could speak the Bard's English, no one would be able to understand.

Heidelberg Man or Woman probably did not wake up one day and start to chatter away to the others in their tribe. It seems extremely doubtful that human speech was developed in this manner. What is more likely is that originally hand gestures were used, followed by mimicking the natural sounds found in the communication of the wildlife creatures which were contemporaneously living at that time.

Slowly, these ancient humans would have started to use a proto-language. That language, as simple as it probably was, would foster social bonding and help make each other understood regarding food sources and mating activity. Language would facilitate and make survival easier.

Putting sounds together into some form that was recognizable to all the tribe must have been overwhelmingly significant. It is too bad there isn't a You Tube video of it. If those first words came from *Homo heidelbergensis*, it is interesting to imagine what the words might be. Possibly the most ancient proto-language was connected to the most basic of all animal needs: finding a good pizza place?

Cooperation for Hunting is Not a Silent Activity

Though there is evidence that *Homo erectus*, and perhaps even on occasion *Homo habilis,* got together to hunt animals, it is Heidelberg Man that shows the first concrete evidence that cooperative hunting was a common tribal activity. One of the indications that helps to identify when and if language was used in some elemental manner, is that *Homo heidelbergensis* was an excellent and successful hunter of big game animals. Bones of creatures like wooly mammoth litter the campsites of Heidelberg Man's firepits.

Some form of speech would have been crucial to convey hunting strategies such as ambushing, trapping, or ensnaring large prey animals. Even if what was needed was to drive mammoths or woolly rhinoceros over cliffs or into swamps, cooperative hunting skills would have been relayed in the form of words between tribal members.

Home Sweet Heidelberg Home

The dwellings of *Homo heidelbergensis* had hearths and special areas which they used specifically for fashioning tools. Every indication is that the living shelters had separate manufacturing areas that were clearly organized. Raw materials were placed in one area, sleeping areas were separate. Multiple huts were arranged around a large outside fire pit, presumably these were used for group feasts. All these living arrangements, which were created and sustained from hunting season to hunting season, indicate sufficient cognitive development and at least a rudimentary language for group cooperation.

Markings on Bone That Need Deciphering

There are bone implements, artifacts that *Homo heidelbergensis* left behind, some of which have several deliberate markings on them. Scientists don't know what these markings were for, but they have been intentionally carved into the bone and were found close to the workshop areas. The incised lines on the bone tools are thought by researchers to have special significance and perhaps have been mnemonic devices.

Were Mom and Dad Equal Partners?

What happens when the band of early hunter-gatherers needs some of the people to hunt for wild animals, yet the vegetables and fruits need to be gathered? Starting very early in hominin history there was probably a gender-based division of labor in their groups. The practicality was obvious: the men could tromp off much more easily and freely than the pregnant and/or nursing women. With much less restrictions placed on them, and their wood spears and stone implements, Heidelberg man followed the large herds while their women gathered. But the men, like all predators hunting swift or dangerous game, did not always come back to the extended family group with a kill. Hunting is an unreliable activity. In the case of Heidelberg man/the hunter, it might have been Heidelberg woman/the gatherer who many times provided for the family.

While the small bands of males were following the trail of the herds, the women were gathering plants and firewood. The grains, seeds, nuts, fruits, roots, eggs, grub worms, insects and dead scavenged animals that were gathered by the women were what could be counted on. There was no means for preserving food. Daily gathering was a more reliable way to get food for the needed daily calories. Generally, it is thought, the women provided the unfailing food source for the entire group.

What Causes Inequality

These were bands of nomadic hunter-gatherers who owned no private property. No one was wealthier than another. Once temporary or semi-permanent shelter was constructed, all other activity was a quest for sustenance. The entire band worked together for the sake of the group. Because of their reason for cooperating, it is likely that men and women had equal influence.

The perception that early hunter-gatherer societies were male-dominated is probably untrue. As a practical matter, the division of labor was based on all the members' efforts counting equally. It is likely that the level of equality carried over into the relations between men and women. Researchers think it was only with the emergence of agriculture, about 10,000 years ago, when people started to accumulate resources, that inequality between the sexes emerged.

Chapter 7: Who Were the Neanderthals? Or What Did Granddad See in Her?

Collectively in pop culture, Neanderthals are often portrayed as big, burly, hairy men who walked around with clubs and when they wanted a date, they would beat the woman over the head and pull her by the hair to a mating spot.

The first discovery of Neanderthal fossils was in 1856 in the Neander Valley in Germany. It came as a shock to the Victorians that there was a species so obviously human and yet not us. Neanderthals are the closest species to humans, and their exact location on the spectrum between humans and other animals has been debated by some scientists since Neanderthal bones were first unearthed.

Finding the Facts

For the first century after their discovery, Neanderthals were imagined as only partially human creatures. More recently their primitive yet human facets have been characterized in cartoons to create a species that is cute and domestic. At present, some researchers now argue that Neanderthals were so close to human that if they walked down the street wearing a suit and a hat, they would go unnoticed. This is highly contested. How much researchers know about Neanderthals will be among the subjects that will be discussed in this chapter.

DNA and Genetic Evidence

Neanderthals or *Homo Neanderthalensis* is an extinct species of humans who lived from approximately 400,000 years ago to 30,000 years ago. Dates for the beginning to the end of this archaic human's existence vary widely, depending on which reference is used.

In modern times, thanks to genomic testing, it is possible to find evidence of Neanderthal DNA in our own genetic code. Uh-oh, how did that happen?

Forty thousand years ago humans were hanging around, looking for similar food sources in the same areas as Neanderthals. Oddly, or not so oddly, depending on how open-minded you are about sexual relationships, the species of Neanderthals met and mated with humans.

Remarkably, some of the Neanderthal/human progeny were able to reproduce. If you remember your animal facts: different species are not supposed to give birth to fertile descendants. Certainly, that is the situation between species such as horses and donkeys (sterile offspring are called mules); polar bears with grizzly bears (sterile offspring are called grolar bears) or zebras with any equine animal (sterile offspring are called zebroids). Theoretically, hybrid animals are the barren offspring of two closely related species of animals. But somehow, this was not always the case for Neanderthals and humans.

I've Got You Under My Skin

There must have been some sex appeal between one species and the other species or it would not be too likely that up to 4% of the DNA of all non-Africans is made up of Neanderthal DNA. As you read this, you might want to think about learning more about your distant cousins. There aren't any photos for your family scrapbook, but there are some very fine pictures of polished skulls you could put into the album, next to the other close group of relatives.

Body Looks

Exactly what the attraction was between Neanderthals and our own species, only the secrets of time possess. The Neanderthals were not only a different species of human, but also very different in looks. Even overlooking the Neanderthals' obvious weak chins, if we were observing a Neanderthal male, (not that any Neanderthals are alive today), we would see a man not taller than 5'5" with his equally short female mate who stood not more than 5' tall. Heavy (170 pounds for the males and 145 pounds for the females), with short legs and flared, barrel-shaped rib-cages, these people would not look like an average mid-westerner, east coaster or even a flamboyant Californian. Their heads were larger than modern humans but formed in a way that might frighten little children (ours, not theirs). Behind massive brow ridges, there was an especially flat braincase. Their long, sloping foreheads culminated with two large eyes and the start of what can only be described as a remarkably big schnozzle. Scientists have had various explanations for the Neanderthal's exceptionally large nose, but to date it is up in the air as to what is the sniff-significance.

Keep Your Chin Up/ If You Have One

To a *Homo sapien*, the Neanderthals would have appeared to have a flawed facial appearance. Neanderthals had a wretchedly weak chin. Humans are used to chins. In fact, no other animals have chins – chimpanzee and ape jaws slant inwards as did Neanderthal's. Between the large Neanderthal proboscis and the diminished chin-line was a mouth that was pushed outward due to protruding teeth. Neanderthals might have been the first poster children for why later *Homo sapiens* invented orthodontics.

Yet, no one should underestimate the strength of Neanderthals. Both the males and the females had thick bones and strong arms and hands. Any hanky-panky that took place with our great-grandparents and a Neanderthal was consensual; that is, at least on the part of the Neanderthals.

Fashion the Neanderthal Way

Neanderthals had hair on their head, but what covered their bodies would be more interesting to look at. The major consensus of opinions at present is that it is possible they had reddish blond fur or fuzz that was the only fashion-covering for their light-colored skin. So far, no evidence for the making of clothing, such as eyed bone needles, has been found at Neanderthal sites. Yet hundreds of flaked stone hide scrapers for removing the flesh of animals have been found. These scrapers could easily have been used to create animal skin-wraps or ponchos for cold weather protection.

In the heat of summer, naked could be considered a fashion statement. Why look fashionable in clothing on a hot day and cover all that nice reddish fuzzy body hair?

Tooling Up

There are vast amounts of Neanderthal functional paraphernalia that have been found at their living sites. When looking at their encampments, Neanderthals left a demonstrable amount of proof of their existence. Besides their tools made of flint and quartz, they also left seashells near their fossilized skeletal remains. There are many Neanderthal hearths that have been discovered, places where they gathered to stay warm and cook their food. Of the many stone tools that have been unearthed are stone pegs and small round rings that might have been created for producing wooden artifacts such as spears.

Neanderthals first appeared in many places in Europe. Eventually they traveled to places like Britain, parts of the Middle East, Uzbekistan and as far as Teshik-Tash in Central Asia. At the time Neanderthals lived in these places, they were habitations of extreme cold. For instance, during most of their existence in Europe, Neanderthals lived very near continental glaciers. Winters were harsh, summers were cold.

The stone tools that Neanderthals created were not too different from their ancestors, *Homo erectus*. Neanderthal stone tools are known as Mousterian, after the site in France where thousands of their artifacts were found. Neanderthal implements have now been discovered throughout Europe, southwestern Asia and North Africa. Though they were skillful and prodigious makers of stone flake tools, they made little technological advances.

Lineages/Ancestry the Old-Fashioned Way

There is no exact consensus among researchers as to when our own species decided the mating game with Neanderthals was over. Or perhaps it was the Neanderthals who broke up with us. One or the other species seemed to have finally figured out that the differences among us wouldn't bode well for a happy relationship. Hopefully no hearts were broken.

Since Neanderthals are not of our own species and we didn't branch off from them, the question remains as to what their lineage is, or, for that matter, what is our own? Most scholars believe that 600,000 years ago, the earliest form of Neanderthal descended from *Homo heidelbergensis* who came from *Homo erectus*. As has already been discussed, *Homo erectus* left Africa and traveled all over the world. *Homo heidelbergensis*, who were living in many European locations, came from this archaic species of *Homo erectus*. The major consensus among researchers is that eventually Neanderthals came from the interbreeding of *Homo erectus* and *Homo heidelbergensis*. Put that in your Ancestry.com app.

The Continentals of Their Time: Neanderthals

Neanderthals never lived in Sub-Saharan Africa. Their kin had set up shop east of the Sahara Desert, which is why Neanderthal DNA does not show up in sub-Saharan Africans living today.

The Climate Was on the Chilly Side

There was nippy weather for most of the time Neanderthals roamed the globe. Thick sheets of ice formed into glaciers. Neanderthals were built for cold weather. Their bodies consolidated heat; their short, stocky stature was an evolutionary adaptation to the harsh climate. And remember their very wide noses? The large nose, it is thought, helped to humidify and warm cold air. It was better than a tiny electric nose blanket.

Evidence of Cold Forcing Neanderthals to Move or Die

Yet, there were places that became too cold and too dry, even for the hearty Neanderthals. In a place called El Salt which is located on a part of a peninsula that is now Spain, Neanderthals who lived in the area until 40,000 years ago diminished in population until finally there is no evidence of them. It is most likely that the change to much colder weather and the diminishing opportunities for food sources caused the extinction of the Neanderthals who lived in this area. Five thousand years later, and in warmer climate conditions, humans began to occupy the same place. What a difference a few thousand years makes.

Diet: the Neanderthal Cook Book

A hundred thousand years ago, Neanderthals were changing their diets. Vegetables such as roots, leaves, forest moss, pine nuts, mushrooms and berries were fine with them, as well as scavenged animals and insects. But as the eras passed they also found time to kill and cook large animals. Carcasses of animal bones litter the Neanderthal living sites. Flexible diets of meat and veggies became favored and made Neanderthals truly hunter-gatherers.

Perhaps their change to a meat and potatoes (in this case turnip) dinner had something to do with their appetites. Neanderthals were hungry people and ate more than *Homo sapiens*: Neanderthal skeletons suggest they consumed 300 to 1,000 more calories per day than a modern human of the same height. Their gourmand tastes and hearty appetites drew them to hunt herds of large game. On the list for Neanderthal barbeques were wild boar, aurochs (a form of ancient cattle), reindeer, red deer, ibex and when times were good, even larger animals.

Up Close and Personal: Coprolites

If you wonder how the scientists of today can tell what Neanderthals were eating, there are researchers who specialize in examining the contents of coprolites. They're called "poop and scoop" scientists because they specialize in studying ancient people's and animal's archaic dung. There is much information that can be gleaned about the eating patterns of the ancients from looking at their fossilized poop. Be sure not to leave any around the house unless you would like future scientists to know what you've been munching.

Neanderthals and Inherited Illnesses

Forty thousand years ago, mixing and mating with Neanderthals provided *Homo sapiens* with some advantages. For instance, Neanderthal blood contained a substance that helped seal wounds quickly and prevented pathogens from entering the wounds. Consider being out in the wild and getting a scrape from the horn of a wooly rhino. The sooner the cut is healed, the better.

Yet this same advantage which was an improvement thousands of years ago is now a handicap for the group of modern humans who inherited this Neanderthal gene. Today we can rely on modern sanitation, hospitals and medical practices to prevent infections. The hyper-coagulation of blood that is found to be a trait in Neanderthal DNA and was a benefit thousands of years ago, is now found to increase risk for stroke, pulmonary embolism, and pregnancy complications.

Through genomic analysis, scientists have found that Neanderthals had several serious genetic illnesses, or at least they risked getting these illnesses. Among the diseases that appear in the Neanderthal DNA are lupus, Crohn's disease, a tendency for depression, type 2 diabetes, and a few other undesirables.

Some of these illnesses, which are diseases that are now associated with modern humans, may have been inherited from our Neanderthal ancestors. The red hair of today, which might also have been inherited from Neanderthals, is beautiful, but it doesn't appear to have been worth the tradeoff.

Empathy and Neanderthals

The climate conditions of cold weather made life for Neanderthals very harsh. Few individuals lived past forty years, and by studying their skeletons, researchers find there were many Neanderthals who endured severe injuries and degenerative diseases.

The disabled and the elderly were not left behind. There are many examples of what can only be interpreted as apparent acts of compassion toward the less fortunate by their healthy compatriots. One such example is a fossilized skeleton from northern Iraq of a male Neanderthal that apparently had received a blow to his head that had fractured his eye socket. He also suffered damage to his collarbone, shoulder blades and had a withered upper arm. At the end of that arm his hand was missing and looked like it had been amputated. He was, in other words, not likely to have been able to forage, let alone hunt. Yet this man lived on for years after his calamitous mishaps. The only way this would have been possible is with the help and kindness from others.

Clearly, Neanderthal's social beliefs in this case were not based on "survival of the fittest", but instead showed a level of empathy and responsiveness that reflected care and consideration for those who suffered physical disabilities.

Neanderthals Buried Their Dead

Intentional burial of the dead is one of the traits of Neanderthals. There is evidence of Neanderthal burials in Israel from the period of 120,000 to 90,000 years ago. Also, in a Neanderthal burial in Uzbekistan, the fossilized body of a 12-year-old boy has been unearthed. There in Uzbekistan archeologists found that the body may have been deliberately "defleshed", perhaps as preparation for interment. There were "offerings" buried with the boy, such as an awl or boring tool made from the horn of mountain goat, as well as other tools. Researchers think that these tools may have had ritualistic significance.

In Iraq there is a Neanderthal burial site that contains many different types of flower pollen which were found with the fossilized body. The flower pollen's existence covering the body gave some researchers the idea that Neanderthals may have placed flower offerings on the grave, much like the actions of modern humans.

Another explanation of the flower pollen that was found in the gravesites is that the flowers were placed on the body due to an archaic knowledge of medicinal treatments.

As with much Paleolithic evidence, scientists do not always agree. The above theories are debated among scientists, some of whom suggest that such ideas "anthropomorphize" this extinct species. The dissenting researchers think that wind carried the pollen onto the gravesite or that the pollen was carried in by burrowing rodents.

Primitive Pharmacopoeia

It is possible that Neanderthals had some ancient knowledge of medicinal remedies. The various types of flower pollen that were found in the Neanderthal Iraqi gravesite are known today to have medicinal effects. The pollen from the *Muscari* genus, for instance, which is commonly called grape hyacinth, is a stimulant and is used as a diuretic. The hollyhock comes from the genus *Althea* and has effects like aspirin, used today to alleviate the pain of toothache. Ragwort from the genus *Senecio* is used as an applicant for preventing bleeding. Woody horsetail or *Ephedra* is used as a remedy to treat coughs and respiratory disorders. And finally, *Achillea*, or yarrow is used in several ways; as a general tonic and as a treatment for dysentery.

There is no way to know if Neanderthals used these flowers as medicinal remedies. But if they did know how to use the flowers for their healing properties, then it is possible that the flowers were placed in the grave as offerings to the dead and possibly as medicines that might help in the afterworld.

Belief in an Afterlife

But did the Neanderthals have a belief in the world beyond death? There is no way of knowing. We do know that primitive tribes existing in recent centuries have concepts and beliefs in an afterlife. Since the Neanderthals are an extinct species and left no written record, a theory of their ideas about the afterlife is completely speculative. No one can prove that Neanderthals believed in a world beyond the living.

Yet there are not too many rational explanations as to why they buried their dead. Nor are their many explanations as to why most Neanderthal burials of the dead have bodies that are found in a fetal position, and many times the bodies are pointing in a straightforward way toward one of the four directions.

The counter-argument is that it is possible that Neanderthals were simply burying their dead because they were not going to put up with a decaying, rotting corpse in their campsite. But the care taken with some of the bodies, the amount of grave goods and the grave sites themselves, including a large, flat rock placed over the body, points to their actions being deliberately more thoughtful than just removing the dead bodies to a garbage heap.

Degree of Cognition/Medical Knowledge

Neanderthals have rightly or wrongly, been described as our not-too-bright cousins who had less than exemplary IQs. But in the grave in Iraq, there is a fossilized body whose hand looks as if it has been amputated. The explanation of stupidity doesn't seem to fit when thinking of what methods would have to be employed to sever a part of an arm with a stone tool and then care for the wound until it healed. Amputation would not have necessarily been difficult for those Neanderthals who were used to butchering animals. But after the amputation, what then? Certainly, it is not possible to leave such a severe wound without further treatment. An injury such as that would require the care and help from another (or several others) to heal.

Astronomical Observations in Burial Practices

In the burial practices of Neanderthals there are some researchers who see indications of the first notions of astronomical observations. Some corpses are oriented in a way that suggests that the four cardinal directions or the motion of the sun or moon were taken into consideration. This is perhaps not as far-fetched as it might sound. The Neanderthals had large brains and were certainly aware of the alternation of night and day, of spring and the other seasons. Winter brought severe weather and the Neanderthals needed to prepare for the hostile conditions. They were hunter-gatherers and the migration of the herd animals was of paramount importance. They had only to look to the sky to see the planetary movements. There is a possibility that some Neanderthals put together an idea to tie their day to day, season to season observations into a big picture explanation.

Art by Neanderthals

The earliest patterned Neanderthal handheld artifact was found in the lower Paleolithic period. An object carved 300,000 years ago from Germany and made of animal bone has symmetrical lines incised into it that have spaces with the line sequences of 7 – 14 – 7. The object is worn, and it is not known what the numerical notations mean, only that the markings are intentional and further demonstrate that Neanderthals had a level of cognition higher than the other animals.

Speak to Me, Neanderthal Cousin

When evolutionary pressures created the genus *Homo*, not only a big brain was added, but also the structure of the brain was changed. These larger brains were far more effective at managing the muscles of the human's slightly larger bodies. Extra neurons were speeding around the bigger brains and needed to be put to good use. What better way than to get the big-brained creatures to talk?

This is not a scientific explanation as to why humans developed speech, but it is true that the prefrontal cortex of the brain of later *Homo* expanded as their foreheads became elevated. And the prefrontal cortex is where we think and make plans.

Couple the thinking brain expansion and the fact that no other animal has a larynx (also known as a voice box) physically low enough to produce sounds as complex as human speech, and an anatomy whose hyoid bone helps to control the tongue, and it seems certain that more than hooting and grunting was taking place between Neanderthals.

The gene that involves early brain development, called the FoxP2 gene, is the gene that is necessary for language development. DNA analysis of the Neanderthal genome has shown that they share this same FoxP2 gene with modern humans. This DNA evidence bolsters the idea that language development was present in Neanderthal children, as it is in our own youngsters.

Neanderthals and Speech

The evidence continues to mount in the scientific community showing that Neanderthals had an ability to talk. Hyoid bones are necessary for speech; and hyoid bones have been found in fossil Neanderthal skeletons. Further, this bone that is essential for speech is found only in *Homo heidelbergensis*, Neanderthals and us.

Because Neanderthals had larger, thicker necks, larger noses and generally were structurally somewhat differently composed in their skulls compared to modern humans, it is thought that Neanderthals spoke, but their voices were quite high pitched. The males may all have been sopranos rather than baritones. Still, if it was possible to speak, there is a possibility they had developed a kind of language. Neanderthal speech would likely have had limited syntax and fewer vowels and consonants due to the restrictive shape of their nasal cavity, which was adapted for living in cold climates. But Neanderthal culture may not have required sophisticated communication.

Researchers know that *Homo sapiens* and Neanderthals met (as evidenced by our Neanderthal DNA). Possibly when these two-species came together, all that was needed was a translator. If only we could have been a fly on the wall of their dwellings, what stories we could have heard.

Neanderthal Demise

Neanderthals existed in Europe for many thousands of years. They were one of the dominant, apex species and hunted and gathered in a kind of non-expansionist limbo. Their numbers never expanded to more than 70,000 at any one time...a number that would have allowed each tribal group enough hunting space to avoid calamitous competition between them. Yet, 30,000 years ago the species of Neanderthals entirely disappeared.

What killed them off is open to debate among scientists. There are several popular theories of Neanderthal extinction.

Extinction Theories:

Climate Change Theory

Climate change disruption and extinction is the theory that because of a severe cold streak in the already cold climate in which they were living, Neanderthals were not able to adapt and died out through natural selection. About 55,000 years ago the climate began to fluctuate wildly from extreme cold conditions to mild cold and back in a matter of decades. Some of the evidence for the theory of climate change and Neanderthal extinction considers the fact that quite a few of their skeletons from 40,000 to 30,000 years ago appear to have suffered from malnutrition or even starvation.

The idea that they died out because of climate change has one large flaw and that is that with their bulkier and more muscular bodies, Neanderthals were well adapted to the cold climate of the Ice Age. Neanderthal bodies were suited for survival– their stocky chests and limbs stored body heat much better than other groups of humans. Further, in the past, their species had undergone several major climatic shifts.

Volcanic Catastrophe Theory

This theory centers on a volcanic explosion called the *Campanian Ignimbrite* eruption. As the unpronounceable name suggests, this was not your ordinary, run-of-the-mill flare-up. Around 40,000 years ago there was a super-eruption in Italy of cataclysmic proportions. The theory is that this volcanic disaster may have played a major role in the fate of Neanderthals. The eruption was the largest seen in Neanderthal times for the past 200,000 years of their existence. It is estimated that the dust and residue from the volcanic explosion strew ash across 1.4 million square miles. The super-eruption would have spread up to 990 million pounds of poisonous sulfur dioxide into the atmosphere. This air pollution would have cooled the Northern Hemisphere, driving down temperatures by 1.8 to 3.6 degrees Fahrenheit for two to three years, enough time to have severe effects on the environment. The Neanderthals were sure to have noticed it. This would have been particularly true since most Neanderthals lived in Europe and Russia.

Since the eruption coincided with the final decline and disappearance of the Neanderthals in Europe, the theory's explanation is that environmental stress on both their bodies and their food supply was the final coup de grace. Many researchers, however, think that the volcanic effects alone are not sufficient to explain the demise of the Neanderthals in Europe.

Interbreeding and Hybridization Theory

This theory is a more pleasant one to examine, having nothing to do with violent destruction. The interbreeding and hybridization theory, as the name implies, suggests that Neanderthals mated with the new apex species (us), a group of *Homo sapiens* that came sauntering into Europe about 40,000 years ago. After modern humans left Africa, they met Neanderthals who were already permanent residents in Europe, and things became friendly and familiar.

The theory is that when Neanderthals and modern humans saw each other, a spark was ignited and for a while it was the era of sex, drugs and rocks and rolling. (Perhaps the last three items are questionable.) But the fact remains that these ancient romps are now visible in our DNA. Genetic analysis indicates that Europeans and Asians obtained 1- 4 per cent of their DNA from Neanderthals.

What this interbreeding theory implies is that our species is the product of hybridization. Though not common in nature, proponents argue that hybridization may allow species to adapt more quickly to new environments. The interbreeding and hybridization theory suggests that mixing with Neanderthals might have helped our own species in forming stronger physiques.

The interbreeding theory has its doubters. One part of the controversy centers on the idea that Neanderthals were "absorbed" into the *Homo sapien* species. The problem is that very small amounts of DNA are found in modern humans. At its uppermost limit, Neanderthal DNA amounts to 4% in modern humans, an amount which indicates that not much interbreeding was taking place. Because there is so little Neanderthal DNA in any modern human, it is not possible to say that the populations merged.

Some researchers have even estimated that the total sexual interactions between *Homo sapiens* and Neanderthals were no more than 70 contacts. Whether that is a mathematical calculation, or some other form of analysis is top secret information.

The other aspect of the theory that is contested is that interbreeding with Neanderthals led to cultural innovations for our own species. At present, there is little evidence to support this idea. Most of the human advances in weapons and improvements in transportation took place thousands of years after the disappearance of Neanderthals. Other than new findings in cave art, most new innovations in human history happened for reasons other than contact and interactions with this different human species.

Displacement and Conflict Theory

If Neanderthal extinction didn't take place through climate change, volcanic eruptions or interbreeding and merging with *Homo sapiens,* how did they vanish? These two different human populations shared Europe for as long as 10,000 years. The Displacement and Conflict Theory implies that *Homo sapiens* drove Neanderthals to extinction. Though Neanderthals had lived in Central and Eastern Europe for hundreds of thousands of years, they were unprepared for a species so like their own who unexpectedly invaded their territory.

The modern humans began to hunt and gather the same food sources as the Neanderthals. And worse for Neanderthals, the new humans were more proficient hunters. They also had better technology and social skills than the Neanderthals. The Displacement and Conflict Theory proposes that competition inevitably arose, and violent encounters were the result.

The Neanderthals were grouped in small tribes whereas the new humans cooperated in much larger groups with better weapons. The Neanderthals lived in troops of no more than several dozen individuals. They hunted together with their close family and friends. Ties between social members were based on intimate contact. Each tribal social structure was probably dominated by one male member who would decide when to fight and when to run. This rigid social structure might not have worked to the Neanderthal's advantage. Depending upon the alpha male's competence and leadership, his decisions might very well have been flawed.

The larger the tribal group, the easier it is to set the terms for the outside groups as to where they will be able to forage and where they will be able to hunt. *Homo sapiens,* who roamed in large groups and were in many instances the immediate neighbors of Neanderthals, might have set the rules of their spatial relationship. The smaller tribes of Neanderthals might have been marginalized and not allowed to gather food, wood or water within certain territorial borders set by the *Homo sapiens*.

The inflexible collective structure of Neanderthals, coupled with the small numbers in each tribe, could prove unfavorable when in a battle with superior opponents who outnumbered them.

In almost all respects *Homo sapiens* 40,000 years ago were no different from modern humans. Though there is little evidence that humans set about an extermination campaign, if they did so it probably would not have looked much different from the ethnic-cleansing scenarios that have taken place in the 20th century. The difference would be that the extinction of Neanderthals would have been based on physical appearance and competition for food rather than religion or nationality.

<u>Weapons with a New Design/Neanderthals Could Not Compete Theory</u>

While modern humans were joining together in cooperating bands of brothers, their ability to use better and more effective weapons for hunting were also improving. The large, rather bulky stone points and blades made from quartzite were finally getting a much-needed upgrade and stones were being whittled down to become smaller and lighter. Modern humans were mounting the smaller weapons in slots carved into wood handles.

The new humans had weapons that could be launched from a distance, an advantage that can't be understated. The farther off the aggressor is from his target, the less jeopardy of personal injury. While the modern humans were throwing spears and standing at a safe distance, waiting for the eventual death of their prey, the Neanderthals were physically brawling with the very animal they hoped to capture and kill.

Neanderthals specialized in stabbing an animal at close quarters with handheld weapons. As the large bodied prey ran to exhaustion they were using their dying breath to fight for their lives. Neanderthals moved in for the kill, plunging their unwieldly weapons into the dying beast at close range and getting badly mauled in the process.

Male Neanderthals' skeletons show that they suffered from the consequences of hunting large animals at close range with heavy handheld weapons. This consequence perhaps explains the reason that Neanderthals had shorter life spans than the humans of the same period. (Generally, Neanderthals died before the age of 40 while humans at that time lived to over the age of 50.)

It is theorized that because of the many advantages of *Homo sapiens*, the Neanderthals were chased from their usual hunting habitats and eventually driven to extinction.

<u>The Disease Theory</u>

Many researchers do not think that any of the above explanations are enough to account for the complete extermination of the Neanderthals. The two species, us and Neanderthals, were closely related. Both species made tools, had fire, were social, and were good hunters...so why aren't the above theories sufficient to explain the demise of Neanderthals?

One problem is that it is particularly puzzling that *Homo sapiens*, the outsider species, were the ones who survived. Before humans arrived, Neanderthals had been in their native hunting grounds for thousands of years and were intimately familiar and knowledgeable about the terrain and the prey animals that they hunted. Also, any natural occurrences such as volcanic poisonous gas in the atmosphere would have affected both species.

The Disease Theory is meant as a clarification to cover the weaknesses in the other theories. It is partially based on historical evidence. Many aboriginal peoples have been completely or greatly diminished by an invading group from a distant land who introduced viral or bacterial pathogens. This has been documented from at least the time of the Spaniards who "discovered" South America and introduced mumps, measles and smallpox into the native populations who lacked all immunity to these foreign, microscopic agents of disease.

There is no evidence of exactly what diseases the new humans 40,000 years ago brought with them, but the pathogens would have been ones that the *Homo sapiens* had built up an immunity to and the Neanderthals had not. The Disease Theory maintains that first contact with the new human settlers may have been devastating to the Neanderthal population and significantly or completely contributed to their extinction.

Dog as Hunter/Exterminator Theory

The idea of dogs playing a part in the extinction of the Neanderthals sounds at first like a bad joke to annoy PETA. Being responsible for the annihilation of our evolutionary cousins is a heavy responsibility to place on man's best friend. But the point of the Dog as Hunter/Exterminator Theory is just that: dogs, or what passed for a dog in days of old, were simply the helpful and useful buddies they have always been to humans. And we may owe more to them than we previously realized. The dog may have to be thanked for helping us to eradicate our Neanderthal rivals.

Renowned for their ability to hunt and chase, and loyal to a fault, dogs could give the modern humans the upper hand when hunting anything from small-sized prey to huge herbivores like mammoths. It is known that dogs were the first animals allowed to live with humans. Other animals such as pigs, cattle and chickens came thousands of years later. The old assumption was that dogs were maintained in camps to guard and sound an alarm, protecting the tribe from invasive animals or enemy humans.

The dogs in the Dogs as Hunters Theory are more accurately described as wolf dogs, not entirely associated with ancient wolves and genetically a long way from the modern hunting dog of today. Forty fossilized skeletons of this distinctive wolf dog have been identified at various sites in Europe. The dog-wolf-anatomy included large and ferocious teeth. It is assumed that the wolf dogs could also pick up scents and chase down any of the animals their masters were hunting.

As early as 40,000 years ago, while taking over new territories, humans could have used some semblance of food reward in the training for wolf dogs. Guided by the hunting humans and staying in packs, the wolf dogs and their humans must have been a formidable sight. What could be more convenient for the humans who then, just as now, hunted with their loyal furry friends, going on long-distant treks to where a large tusked hairy elephant had been cornered? There, with superior weapons, the humans could take over and finish off the hunted prey. Imagine a scene like riding with the hounds, but without horses and a fox the size of a mammoth.

The Dog as Hunter/Exterminator Theory describes Neanderthals as collateral damage, not being targeted by humans, but simply being out-hunted. We couldn't do it without the dog. Or at least, this is the theory.

Too Compassionate for Their Own Good Theory

In recent times many researchers have stressed the compassion and kindness of the Neanderthals. There is evidence of Neanderthals exercising exceptional consideration and kindheartedness, at least to members of their own group, as they nurtured the old and buried their dead. There is a possible explanatory analogy as to how the Neanderthal extinction took place. The dodo bird is a species which vanished in the late 17th century. It is believed that the dodo went extinct because Dutch sailors ate the animals into annihilation after finding that the bird was incredibly easy to catch due to the fact it had no fear of humans. Compassion coupled with a certain cluelessness, it is theorized, could have done away with the Neanderthals.

Pick Your Theory

All the above theories have their scientific supporters. Some researchers are sure that only one theory is valid, while others believe that the extinction of the Neanderthals has multiple explanations. What is not disputed is that Neanderthals, who had lived in Europe for over 300,000 years and had been on the planet in various locations around the world for well over 400,000 years, disappeared within a few thousand years of our arrival.

Chapter 8: Is It True What They Say About Tiny People? /*Homo naledi* and *Homo floresiensis*

<u>*Homo naledi:* The Small South Africans of Long Ago</u>

There were other ancient human species that lived and thrived at the same time as our own species. Also, not all *Homo sapiens* left their ancestral home of Africa. Some stayed, maybe because they liked their neighbors. The *Homo sapiens* who stayed in Africa lived in a neighborhood filled with strange looking hominins which researchers have only recently discovered and identified. They are named *Homo naledi* and they existed until recent times. *Homo naledi* were still hanging out in South Africa around 200,000 years ago.

<u>A Different Kind of Look: Small Humans in South Africa</u>

Homo naledi had a small skull and sloping shoulders like an ape's. Yet, some of their other features such as the shape (as opposed to the size) of their skull, their face, and their teeth, are modern enough that they are placed in the genus *Homo*. The *Homo naledi* skeletons proved difficult for researchers to classify. But even at the time they were alive, they might have been quite confusing to identify. Perhaps *Homo sapiens*, on first coming across them, might have thought, "Should we eat these guys or make friends with them?" There are no written records of what an eyewitness to their first encounter might have overheard. One researcher has called *Homo nadeli's* body, "bizarrely primitive". Their torso was shaped like a small ape's and with their tiny-brains it is clear that, except for a few of our modern politicians, they didn't resemble our own lineage.

There are many *Homo naledi* skeletons in caverns in the Rising Star cave system in South Africa. An impressive discovery of hominin remains has been found at Rising Star—the single most productive fossil site of its kind in Africa. To date, among the two main chambers, over 1,630 specimens of *Homo naledi* have been discovered. One example of the chamber finds, with almost complete skeletal remains, are bones that belong to three individuals—two adults and a child. The most intact adult skeleton nicknamed "Neo" is represented by bones of the head and body and was probably male.

Technically Dating the Finds

Many different dating techniques were used to determine the age of these skeletons. Layers of calcite had been laid down ("flowed over") by mineral-rich running water and had covered some of the *Homo naledi* specimens. These flowstones were radiometrically dated. Also, though it entailed destroying a bit of sample fossil, the scientists ground up a small amount of the teeth, and then exposed the sample to U-series (Uranium series dating) and ESR (electron spin resonance dating), which allowed researchers to establish the dates when the specimen skeletons were alive (335,000 to 226,000 years ago).

A Mystery: Why Are the Bones There?

One of the most fascinating things about this small-brained species, who seem like relics mimicking much earlier extinct orders of hominins, is where their bones are located. The scientists have had to squeeze through slots less than a foot wide to enter the cave excavation area. Only the young, small and agile paleoarcheologists needed to apply for onsite work. What is more baffling than a tiny scientist working in such tight quarters is why the bodies of *Homo naledi* are there in the first place.

From the cave entrance to where the bones are deposited is a narrow passage that is dark and requires crawling and vertical climbing. Researchers are sure the entrance has been a challenging access for many thousands upon thousands of years. None of the bodies ended up in the bone chamber at the end of their lives by accident. Because the fossilized bones of this species have been found in several almost inaccessible chambers within the cave system, with no possibility that these hominins were camping out, researchers have speculated that *Homo naledi* might have been burying their dead. If the skeletons found at Rising Star are examples of some form of purposefully disposing of the dead, other than digging holes and burying their dead, then this would mean that the mortuary practices of Neanderthals and of modern humans were not the first examples.

The evidence for suggesting that *Homo naledi* might have been motivated to intern their dead is that the multitude of bones found in the chambers are almost exclusively the remains of *Homo naledi*. Also, there are no alternate entrances into the cave chambers. Nor would so many *Homo naledi* be in the cave chambers roasting chicken legs. Not only are the entrances to the cave chambers exceptionally diminutive, but the chambers themselves are small and the bones are piled together. It would be an impossible place to live, and inconceivable that all these individuals ended up there by coincidence.

Critics Insist That There Are Only Two Species Who Bury Their Dead

Yet, there are critics of this idea. Death rituals have been ascribed only to Neanderthals and modern humans. Many cannot accept the possibility that primitive, small-brained *Homo naledi* could have engaged in the deliberate disposal of their dead. It has been assumed that the behavior is embedded only in our genetically close cousins the Neanderthals, and us. If this behavior appeared before humans, it threatens to dissolve the distinctive boundary between us and other animals. Bereavement rituals have been viewed as solid evidence in tracing the emergence of human uniqueness. Though behavior itself can only be conjectured, funeral practices and the bones that have been buried, are proof- positive of the actions of humans. Other than funerary practices, determining exactly what makes us essentially human and distinct from other species is a more difficult task than it would appear.

The Controversy Continues: A Story of a Puzzling Situation

Questions abound. One problem is that *Homo naledi* had a brain the size of a gorilla and did not show signs of great cognition. How could a creature with a brain much smaller than the size of a modern human's have reasons other than possibly waste removal, to bury their dead? There are no tools of any kind that have been found in the cave chambers, so no rituals centered on digging a hole to contain remains could be the reason for the appearance of the bones in the caverns.

The entrances to the chambers could only be reached with some sort of fire to light the way. Control of fire would have to be ascribed to this species and no implements such as primitive torches have been found at Rising Star. But no matter how it was accomplished, *Homo naledi* went through great effort to carry their dead through darkness, far into the cave.

Most scientist are waiting for a more extensive search and peer review of the Rising Star cave chambers. Researchers estimate that less than five percent of the chambers in the cave system have been unearthed. Still, the reality is that accumulations of multiple bodies of all ages of a new species of hominin have been discovered all together in a cave system in South Africa. And the bodies look suspiciously as if they have been deposited in those chambers with thoughtful determination. Patterns of grief for the loss of family members are not limited to humans. Elephants have been observed grieving over their dead. Other than burial practice, so far there have been no other explanations to account for these findings.

How Unique Are Homo sapiens?

By challenging accepted ideas, it is possible to see that some of the uniqueness credited to *Homo sapiens* is an oversimplification. With the discovery of *Homo naledi*, it is possible that the hierarchical diagram of humans as the apex species which have the exclusive claim to signature behaviors such as burials of their dead may be another human trait that will need to be expanded.

Homo floresiensis: AKA "The Hobbits"

Fables and myths are ways we have of explaining some phenomenon that seems otherwise inexplicable. One common legend that is thousands of years old has to do with the existence of "little people." Surprisingly, this myth takes on fresh meaning when learning about the paleontological digs on the Island of Flores. Although little people are common characters in ancient legends, no one would have thought that any might have existed on Flores, which is part of the Lesser Sunda Islands in Indonesia.

In the early 21st century, researchers began excavations at a large limestone cave called Liang Bua on the Island of Flores. They found a hominin arm bone and tooth and skull located about 29 feet below the cave floor. As they continued to excavate, several other partial skeletons were uncovered.

Flores Was Always an Island

Flores is a large island that during the last many millions of years, was never connected to the mainlands of Australia or Asia. The flora and fauna are quite distinctive. A million years ago, the hominins who reached the Island of Flores would have had to go by water. Even during the Pleistocene period (2.6 million years ago to 11,700 years ago) when 30% of the land area of the earth was covered by glacial ice and the ocean's water level was low, the Island of Flores was just that: an island.

There were, however, humans who made it to Flores. Scientists think the earliest migrants, who were *Homo erectus*, made their way across the water in rafts from the East Indies. Stone tools have been discovered that date to 700,000 years ago, suggesting that archaic human ancestors had migrated to the islands of Indonesia long before the evolution of modern humans about 350,000 to 200,000 years ago.

Very Strange Looking Skeletons Were Found

"Hobbits" is the nickname of the bones of the hominin remains that have been found. *Homo floresiensis* is the scientific name denoting this strange group of humans who lived on the Island of Flores. There are many debates about whether the skeletons found on Flores represent an ancient population that was highly different from other human populations. The dating of these small people is still being studied, but it looks as if they lived on Flores from about 150,000 years ago to 45,000 years ago.

The fossil skeletons that have been found on Flores are archeologically quite different from skeletons of other hominins. Despite their nickname, they are not the long-lost little people in J.R.R. Tolkien's books. And they didn't inhabit the "lands of Middle Earth" as did the hobbits in the Tolkien stories. But it must be admitted that there are some parallel similarities about Tolkien's fictional account of his descriptions of the little people and the archaic dwarf-like humans who occupied the Island of Flores. In appearance, *Homo floresiensis* rather incredibly, resemble Tolkien's fictionalized creatures. Like the imaginary beings of "The Hobbits," *Homo floresiensis* also had small heads, small bodies and proportionally very large feet. Archeologists are still wondering what Tolkien knew that they are just finding out?

What the Hobbits Looked Like

The most complete skeleton that has been discovered was an individual, about 3 feet 6 inches tall and weighing about 66 pounds. It looked to the paleoarcheologists like a female child until the teeth were examined. The jaw contained wisdom teeth that were fully exposed and demonstrated signs of wear. What the scientists were forced to conclude is that this was a tiny adult, which was about 30 years old.

There is controversy among scientists as to whether this species should be part of the taxonomic tribe called *hominins*, comprising modern humans and their ancestors. Besides the tiny body, *Homo floresiensis* had very small eyes and lacked a chin. The presence of a chin is one of the defining traits of *Homo sapiens*.

A Human and Hobbit Overlap?

On Flores, fragments of 12 other individual skeletons were recovered that were associated with the most complete skeleton, as were stone tools. In addition, charcoal was found, suggesting the use of fire. The initial dates of this species overlap with the time modern humans are known to have arrived in Indonesia. About 45,000 years ago humans were traveling through the Indonesian islands. The question is: could modern humans have co-inhabited the island with the Hobbits of Flores? This question has yet to be answered.

Hobbits and the Microcephaly Disorder

Another of the controversies is that *Homo floresiensis'* brain is about the size of a chimpanzee's. The conventional thinking is that human evolution advanced from upright walking apes which progressively evolved to larger bodies and larger brains. Because of the exceedingly small brains of the Hobbits, many anthropologists think that the entire species of *Homo floresiensis* may have had a developmental or pathological disorder.

An explanation from some researchers about these tiny hominins with their tiny brains that evolved and survived into the time of modern humans', is that the Hobbits were suffering from a condition known as microcephaly. Microcephaly is a condition which is characterized by people who don't develop fully and have very small brains and heads and usually diminished mental capacity.

Hobbits Were Capable of Many Things

There is evidence that the Hobbits used and controlled fire. They also created stone tools. Evidence exists that they killed and butchered the now extinct pygmy elephants that lived on the island…presumably the pigmy elephants were slaughtered as a source of food. Pigmy elephants were about the size of a cow and would not have been an easy catch for one person who was shorter than 4 feet tall. This implies that the Hobbits went on group hunts. Most paleoanthropologists think gathering groups together to cooperate in hunting expeditions would require some form of advanced cognition. All of which would seem to undermine the theory that the Hobbits had excessively reduced mental capacity.

Islands Can be Stranger Than You Think

Islands are places where anthropologists have found evidence of both gigantism and dwarfism. Over generations, sometimes island animals change dramatically. Growing large on an island may be due to lack of predators. Shrinking in size on an island in comparison to earlier ancestors may be due to limited resources and a reduction in territory; e.g. the smaller the species, the less food that is required.

Below are examples of the strange effects of the Island of Flores on some of the extinct or extant creatures of Flores:

Giant rats (still extant) about twice the size of the average brown rat;
Giant storks, (extinct) that were 5' 9" tall and weighed up to 35 pounds;

Komodo dragons, (still extant) the largest member of the lizard family that grows up to 10 feet and 150 pounds;
Pigmy elephants (extinct) a dwarf elephant that grew to be about 5'5" tall.

The Theory of Island Dwarfism and *Homo erectus*

There are many who think that the Hobbits of Flores evolved from a population of ancient *Homo erectus*. *Homo erectus* left Africa and reached Indonesia about 1.3 million years ago. These are the first descendants that really looked human and there is evidence that these archaic hominins were in the Indonesian archipelago; stone tools that were typical of *Homo erectus* have been found on the Island of Flores. This theory maintains that *Homo floresiensis* evolved from *Homo erectus* and adapted to their surroundings. Their diminutive size is an adaptation of island dwarfism.

The critics of this theory point to the fact that dwarfism is not usually concurrent with a reduction in brain size, which is the case with *Homo floresiensis*.

Extinction of the Hobbits

There are several possible explanations for the extinction of the Hobbits of Flores. Volcanic disturbance on Indonesian islands is dangerous and many times spews out a more violent form of eruption than other places in the world with volcanic activity. The hot flows of gas and rock can reach speeds of 500 miles per hour. The fossil finds of the Hobbits, giant storks, and pigmy elephants, (all of them extinct), were found below a thick layer of volcanic ash. This could mean that the Hobbits were the victims of naturally occurring forces: volcanoes specific to the Island of Flores, which destroyed them and many of the other island species.

But then, there is a possibility that their extinction was not the result of Mother Nature and her destructive forces. Another theory is that although the Hobbits may have lived on the Island of Flores for over 100,000 years, some of the paths of dispersal of modern humans are routes that flowed into Indonesia. If Hobbits coexisted along with humans, it might have been a tempestuous relationship. The Hobbits were little, and their stone tool kits were not advanced. There is a possibility that our species out-competed these small-brained hominins. According to this theory, the Hobbits were driven to extinction because of the loss of their territory, food sources, and possibly the problem of being used as target practice.

Hobbits represent a unique population of some form of human that were isolated for thousands of years from the rest of humanity. The discussion around the Hobbit's identity and classification, along with questions about their way of life, will most probably be a lingering debate. If only Tolkien were still alive to answer more archeological questions.

Chapter 9: More Origin Ancestors: Denisovans

You might wonder what your DNA contains. How many species do you come from? By now you might realize that you are not "pure". There has been mingling among species. Perhaps you have become reconciled to the fact that a tiny bit of Neanderthal DNA comprises some part of your chromosomes. You might be surprised to learn that there could be more hitchhikers on your double-strand DNA helix.

Meet the Denisovans. Denisovans are a mysterious population of hominins that lived at the same time as the Neanderthals and *Homo sapiens*.

Before you prepare to make a space for a separate section in your family scrapbook, you should know that not all modern humans are biologically related to the Denisovans. Denisovan DNA of up to 6 percent is present in the chromosomes of some modern South Asians, Melanesians and Aboriginal Australians, as well as others. Just as in the case of Neanderthals, the Denisovans are a separate species who at one time in the distant past, mated with *Homo sapiens*. And some of their offspring were fertile.

The definition of a "species" is a group of organisms that can interbreed and produce fertile offspring. As explained elsewhere in this book, theoretically *Homo sapiens* (us) might mate with an alien species, but our offspring would not be fertile. Yet, strangely enough, though the Denisovans are not our species, their progeny is still dimly represented in some modern humans today. How this happened can only be imagined. The most likely scenario is that after some frivolous sexual behavior where no one was remembering the genetic divide, these two-different species produced offspring in the form of a few broods of children capable of reproducing.

<u>Three Different Types of Settlers</u>

When our *Homo sapien* ancestors first migrated out of Africa around 70,000 years ago, they seemed to have met buddies: Neanderthals left North Africa about 400,000 years ago and settled in Europe and parts of western Asia. Though not much is known about the Denisovans, it is possible that they are a much more recent addition to the human family tree.

A Species Story Told Minus its Artifacts and Most of its Fossil Remains

In 2010, near the borders of Russia, Mongolia, and China, in a cave in Siberia, a remarkable find was discovered. Paleoanthropologists uncovered a 40,000-year-old tooth (later two more teeth were found) and a fossilized pinkie bone. The tiny bone was no bigger than a dime. Yet, scientists were able to extract nuclear DNA from the pinkie bone. The genome of both modern humans and Neanderthals are complete and a comparison study was undertaken with the DNA found in the pinkie bone. From the genetic traces of the pinkie bone, genomes were matched and contrasted.

The ability of researchers to identify a specimen from a small amount of bone analysis is a new and remarkable achievement. Never in paleoarcheology have extinct humans been detected and described from their genome: no fossil skulls or stone artifacts were necessary. Even now, several years after the extraordinary discovery, nothing is known about what the Denisovan faces looked like, how they lived, what tools they used, their height and weight, or if they buried their dead. All else is lying in wait under the vast surface of the earth, available someday to some paleoarcheologists lucky enough to locate and excavate the sites.

Outstanding Discovery

The genetic pinkie bone analysis revealed significant information. The outcome from the DNA tests is that the 40,000-year-old pinkie bone was from a small female child. The little girl was closely related to Neanderthals and to some modern humans. This discovery was extensively hailed in the scientific community because anthropologists had discovered a new species of archaic hominin. They are called "Denisovan" after the Siberian cave where the pinkie bone was discovered. DNA analysis has become so sophisticated that the researchers also have been able to deduce that the young girl had brown hair, brown eyes, and brown skin.

Between the recent discovery of the tiny part of a small girl's finger and the separate discovery of the molars in the same Siberian cave, coupled with the amazing advancements in DNA analysis, it is likely that there will be other bones and other undiscovered hominin groups that are waiting to be revealed. Be prepared to be astonished.

That Certain Magnetism Making Denisovans Irresistible

That partial finger bone was the first evidence of the Denisovans, which is now thought to be a distinct branch of the genus *Homo*. The Denisovans must have had something special that made them particularly attractive and tempting; perhaps it was curly (or straight) brown hair or a certain swagger or sway, or maybe (as in the song sung by Marilyn Monroe) it was their "personality"? DNA analysis doesn't lie, and the cat is now out of the bag: both Neanderthals *and* modern humans mated with Denisovans during the past 100,000 years. If only we knew their alluring secret.

Possibly the mating urges were brought on by the desire to breed with a mysterious species that also had fascinating work objects. Buried in the same excavation layers as the Denisovan fossils were many kinds of different artifacts. The technological skill of the stone tools varied from the crude points of Neanderthals to the slender stone blades and bone points typical of modern humans. It is not possible to know just what group created these advanced tools; the cave has been visited by archaic humans and others. But if some of the tools were fashioned by the Denisovans, think of how appealing it would be to mate with one of them: a group who may have roamed vast expanses of Asia with tools as sophisticated as those made by humans. Well, perhaps that doesn't sound enticing enough to you, but you didn't live thousands of years ago. At that time, who would find Bill Gates so appealing?

The Denisova Cave: A Scramble of Bones, Tools and Other Artifacts

Denisovans are genetically like Neanderthals and modern humans. Researchers presume they could create proficient stone tools. But the extent of Denisovan tool making technology is difficult to ferret out because for 100,000 years various humans have frequented the cave in Siberia. Archaeologists can only speculate about who made what type of tools. It seems that the Denisova Cave was a cozy rest place for many: Denisovans, Neanderthals and modern humans. All their fossils and stone artifacts are in that cave chamber of bones. Added to that is the problem that through the thousands of years hyenas and friends have been chomping on many of the bones in the cave. It's difficult to sort things out. Many more research hours will undoubtedly go into the organizing and classification of the archaic artifacts and fossils. What is certain is that the hyenas can be written off as creators of the stone tools.

A Papa in Papua

There is no doubt that there was species mixing when the Neanderthals, *Homo sapiens* and Denisovans met. But how did just a small smidgen of Denisovan DNA get all the way to New Guinea? Not to paint too explicit a picture but, on their long journey to see the world, modern humans passed through eastern Eurasia and were smitten with some of the Denisovans who resided in Siberia. One would hope that an appropriate period ensued and that no hearts were broken, then the ocean was crossed and 45,000 years ago these *Homo sapiens* with their Denisovan DNA, settled in places like Papua, New Guinea. These were the ancestors of Melanesians. Today, 3 to 5 percent of DNA for Melanesians is Denisovan. The fact that the species of Denisovans were discovered in a cave in Siberia, but the genomes of modern humans living in Southeast Asia have some Denisovan in their chromosomal make-up, suggests that at one-time humans and perhaps Denisovans ranged widely across Asia.

The First Yachts Were Rafts

There are no boats or any kind of floating craft that have been preserved from so long ago. But no matter how much ice was locked on the land in the form of glaciers, some places in the world would have been impossible to reach without using watercraft to cross the sea. Since no one is suggesting that Denisovans or any archaic hominins walked on water, it seems reasonable to think that they must have used watercraft. It is speculated that some of the earliest humans arrived in New Guinea by boat. Many researchers believe that seaworthy boats were built as early as 600,000 to one million years ago. If this is so, *Homo erectus* and/or Denisovans may have been confident on both terra firma and the great waters of the world.

Using plants and stones is not a new story. This is a book about people who were using very little else for their survival. Available vegetation and rocks are the identifying markers of the Stone Age. Not too surprisingly, theories about Stone Age boats have to do with these key components. The earliest boats are believed to have been rafts, created out of bamboo or reeds. As stone tools became more technically advanced, it is believed that dug-out canoes were crafted out of long logs.

The raw materials to create a seaworthy boat would have been anything that floated. Ancient people could lash together bamboo, wood logs, or reeds, tied together with vines or palm fibers. Starting more than 1 million years ago there were stone cutting tools that could handle the job of slashing reeds, bamboo or palms and vines. It would not take a very complex cutting tool to fashion a simple floating contrivance.

Perhaps plank/reed/bamboo rafts were first used as fishing platforms. As the platforms floated for long periods of time, they then began to be utilized as boats to be navigated longer distances. All that was needed to navigate was push poles in rivers, or oars for ocean travel.

Having No Preconceived Ideas Proved Remarkably Effective

From the distances traveled, it looks as if early types of hominins were anxious to get to the other side of the world. And since there were no nonsensical rumors such as wondering if the earth were flat, these ancestral cousins probably had no thought that if they went too far, they would be falling off the globe. There might be some benefits to not over-thinking a situation.

Denisovans Were a Hearty Bunch

Researchers have credited Neanderthals with being the genetic providers for modern human adaptation to severe climatic conditions. Since the findings of the Denisovan evidence, it appears that credit needs to be shared between Neanderthals and Denisovans. It might be that these elusive early humans gave modern humans a genetic way to survive in the farthest reaches of the earth where climate conditions are punishing.

Since Denisovan DNA is found in the genes of many of the heartiest living humans such as the Inuit people from Greenland and native Tibetans, it is thought that it might have been the healthy, vigorous Denisovan part of their genetic makeup that helps these modern humans live in extreme cold and high altitudes. Certainly, they are genetically adjusted to their edgy environments. One of the genes which regulates blood hemoglobin and another which is related to the amount of body fat distribution are thought to be inherited from the obscure ancient species of Denisovans.

Possibly an Eastern Variant of Neanderthals

Even in families of today, heritage is sometimes questionable, and since so little is known about the Denisovans other than traces of bone and DNA, researchers are not in agreement as to their origination and relations. The Denisovans are not orphans who no one claims. It is rather the reverse. The gene pool in people living on islands in Southeast Asia and Oceania, some indigenous Filipino groups, the people of Papua, New Guinea, and Aborigines from Australia, as well as a few mainland East Asians, are all connected by having some Denisovan DNA. There are undoubtedly many people who would match up as being distantly related to this ancient species. Thousands, if not hundreds of thousands of people living today might be willing to call Denisovans their distant relatives to the nth degree. Despite the popularity, so far Denisovan ancestral heritage is controversial and remains somewhat of a mystery.

Another Neanderthal by a Different Name

Neanderthals genes have a relatively high percentage of Denisovan genetic heritage. There is a theory that Neanderthals who lived in Eurasia might have split off with another group who spread to Asia and that those groups might be one in the same. The group who stayed in Europe we call Neanderthals, and the group who spread to Asia we call Denisovans.

Being isolated from each other through a hundred thousand years would create physical differences. According to this recent theory, Denisovans might be the eastern variants of Neanderthals. This would make them more than kissing cousins; this would make them brothers and sisters.

More Distant and Even Farther Away: Relatives from the Past

As Alice in Wonderland would have said, things about the Denisovans have become "curiouser and curiouser". The plot thickens from a mystery of lineage that includes two other species (us and the Neanderthals) to a genetic marker that shows that Denisovans were interbreeding with another unidentified species. According to researchers who study paleogenetics, that unidentified species is much, much older than any of the other modern human species. Somehow, that archaic hominin species, which scientists currently are unable to identify or give a name to, was living alongside the Denisovans. By extension, the unknown species might have been living alongside our direct kin. The Paleolithic era must have been crowded at times, and considering the variants of species, "normal looking" might have appeared very differently than the way we understand it, even if we compare it with the strange looking people we sometimes see hanging around in international airports.

To Sum Up the Cousins/Ancestors/Siblings of *Homo sapiens*

As we leave our archaic past and start to explore our more recent brother and sister *Homo sapiens*, let's review our family bush. We'll scuttle into the brambles for a brief look at what we are leaving behind: climbing around on the many stems reminds us that humans were around long before history. Millions of years ago, significant events of the hominin kind were taking place without record keepers. Today we have researchers who have delved deeply into the archaic undergrowth; they are called paleoanthropologist and paleoarcheologists.

Most of the earliest prehistoric hominins were insignificant and were not too different than the animals surrounding them. Though there had been a split in the simian family 7 million years ago, our first bipedal cousins (*Australopithecus*) didn't look too different from other great apes. But walking upright millions of years ago was a bigger deal than it looked. Binary legs to walk on were one of the causes of our genus *Homo* breaking off from our chimpanzee and other great ape relatives. Slowly *Australopithecus* climbed down from the trees which grew by the rivers and lakes and walked with two feet into the warmer savannah. As time went on, features changed, including arm length, adaptable hands and feet, mouths, throats, and diet.

The first proto-human predecessors connected to us appeared on the scene 3 million years ago. A few hundred thousand years ago, the parents to all the humans now on earth called Mitochondrial Eve and Chromosomal Adam appeared (though not necessarily in the same period). There were several versions of hominins which at that time were roaming the planet, probably on the move to seek better food sources and warmer environments. These archaic hominins were deeply integrated into their natural surroundings. Their evolutionary motivation was a compulsory desire to endure. But though their new methods were more involuntary accidents than cognitive breakthroughs, the result was that there were fresh approaches for survival.

Other physical adaptations were happening to the early hominins, the result of natural selection. Hairless skin had advantages such as not spending hours picking mites and other insects out of fur. The color of their epidermis, changing to meet their climatic situation, morphed so that hominin skin colors were forever altered. Smaller teeth, weaker jaws and shrinking intestines came about as fire changed food into mushier substances. Calories were needed by our ancient ancestors and cooked vegetables and meats made it possible to eat enormous amounts of energy-dense foods in shorter periods of time. Accumulating body fat, the only reliable way to store food before the invention of fast food restaurants, was also assisted by cooking.

A massive expansion of brain size was taking place. From 2.5 million years ago up to a few thousand years ago, the human brain has tripled in size. This and other new human adaptations were going to revolutionize the genus *Homo* from an insignificant bipedal, timid creature, to the apex predator of the globe. But all this was an incrementally slow process. The vast time it took for the developmental progress of the genus *Homo* up and until now should not be underestimated. Like the glaciers which grew and retreated, so too, did the archaic hominins whose various species life spanned thousands to millions of years here on earth.

The first tool hominins created was fashioned from the abundant rocks that were under their feet. Over a million years ago *Homo habilis* figured out how to shape rocks to be used for butchering and scraping. Later humans like *Homo erectus, Homo ergaster* and *Homo heidelbergensis* raised the craft of the Acheulean handaxe to a fully formed multi-useful tool. The weapon with the longest shelf-life for humans has not been the gun but an adaptation of a triangular sculpted rock we call an axe.

Homo habilis is credited with being the first archaic ancestor to develop pair bonding. Pair bonding, an adaptation which sounds like an excuse to imprison men into a lifetime of domestic servitude by marriage-obsessed-single-women, was critically important for the survival of the species.

The human's upright walking position had anatomically changed the location of the legs, which were now directly under their hips. This created a narrow birth canal for the human female. Also, due to expanded brains, the baby's skull became much larger in size. The female no longer could easily give birth to the baby. With the adult female's legs directly under her pelvis, and the baby's large head, the child at birth now needed to be delivered prematurely. The baby was not going to learn to walk for over a year and the lactating mother and her offspring needed protection.

Pair bonding meant that the male would stay in the company of the birthing female and assist in childrearing and supplying food during the crucial several years of the child's development. Without the paternal male's help during childbirth and his provisional help and protection during the child's early development, the hominin species would not have been able to survive.

Though some groups of the genus *Homo* left their East African surroundings, evolution did not cease its fecundity. Numerous new species developed after or during the time of *Homo erectus*: *Homo rudolfensis, Homo ergaster, Homo heidelbergensis,* Neanderthals, *Homo naledi, Homo floresiensis, Denisovans,* and *Homo sapiens*. Later humans showed signs of significant new adaptations and development.

Homo heidelbergensis is usually thought to be the species whose anatomy changed enough to allow for complex speech. Heidelberg Man is the species related to both modern humans and Neanderthals. For ancient humans, as soon as speech became anatomically possible, as it appears to have been for *Homo heidelbergensis*, language became advantageous for purposes of cooperation in hunting and the sharing of knowledge.

Many of these ancient humans existed at the same time. Putting them in separate chapters and making lists with one human species over or under another is simplifying their existence and positions on the human timeline to the point of equivocation.

Homo sapiens evolved in East Africa starting from 350,00 to 200,000 years ago. There were various hominins on the planet at the same time. Our ancestors mated with some of them. Some of them they apparently did not. But regardless of how many species interbred with humans, genetic evidence continually reveals that there has never been a "pure" *Sapien* species.

Thousands of years ago anatomically modern humans were shuffling around the landscape and they met many unusual hominin members along the way. In Europe and parts of Asia there were Neanderthals. Also, in the neighborhood, particularly in Asia, there were Denisovans. In isolated pockets there were unidentified ancestors, possibly a very archaic species of *Homo erectus.*

On the Island of Flores, the species of *Homo floresiensis* (the Hobbits of Flores) were taking their tiny steps to match their miniature body size. And *Homo naledi,* another species of diminutive size, were active in caves in South Africa doing what some researchers believe to be burying their dead.

The pantheon of discovery of archaic humans is likely not over. Look for new findings within our lifetimes. To quote Alice one more time, things are becoming "Curiouser and Curiouser".

Part Two: Modern Humans

Chapter 10: It's A Humankind – Kind - of Thing: *Homo sapiens*

The Origin of Humans/*Homo erectus/Homo sapiens* and Darwin's Theory

Far from being the superior beings that some humans think our species is, the evolution of what became us began insignificantly and without any obvious purposeful aim. We call ourselves "*Homo sapiens sapiens*" which means "wise, sensible and judicious" (it seems we can't have too many "sapiens" attached to our title). But rather than being the chosen preeminent species, we exist mostly because of chance events. If there hadn't been a catastrophic incident 65 million years ago, the reptiles would most probably be the kings of the earth and if we existed at all, we'd be limited to small, protected nature preserves. Picture the tribe in the film "King Kong" who built gigantic bars around their village to keep out the prehistoric monster (and who offered up good-looking virgins for Kong's erotic appetite).

The Little Species That Could

In trying to answer the rhetorical question, "Were we meant to dominate the planet?" we need to look back at our embryonic history. After reading Part One of this book it would have to be said that considering the beginnings of archaic hominins three million years ago, it didn't look like a winning situation. Our ancestors were timidly trying-out bipedal walking on the savannah, but only in the daytime hours. At night they took shelter in the trees for protection. A comfortable nighttime habitat at the time of *Australopithecus* was how high and wide the branch was on which to sleep. The circumstances are still unknown as to what led some of those *Australopithecus* to strike out and walk down the road toward humanity. But just like an unplanned pregnancy, their steps led to many unintended consequences.

By 250,000 Years Ago

Leaving aside many eras, the question is: over 250,000 years ago, would there be anything in the mannerisms or traits of the very early specimens of the species *Homo sapiens* that would indicate that they were going to be keepers? Probably not. They hadn't produced many off-springs. They were a group of not numerous, not large mammals who looked, by all appearances, rather helpless. Nor were they creating exceptionally advanced stone tools. Instead of their clever hunting and foraging skills, the climate was the major deciding factor in whether they ate or starved.

And there was competition in other parts of the globe. Neanderthals had come before *Homo sapiens* and had created a cozy niche for themselves in Europe and parts north. *Homo erectus* was a heavily invested species in parts of Eurasia. If primordial Las Vegas gamblers had been around 250,000 years ago, they would not have given *Homo sapiens* great odds at survival. But of course, they would have been wrong. Our species left Africa and became, in time, the apex predator it is today. Having colonized and taken over every part of our own planet, futurists are now eying other planets as possible Earth Two.

How We Did It

Three million years ago, the *Australopithecines* had evolved into two distinct groups, the gracile and the robust forms, as well as many variants. Some scientists call these species "the walking apes" though to be kind, at least *Australopithecus* females would probably have thought better of themselves. The female

Australopithecus might well have been the selector of the more superior males in their group. They wanted a guy who was taller, cuter and had a good bipedal walk... cherchez la femme.

By 2.3 million years ago the first of the genus *Homo* appeared: *Homo habilis*. The research is not clear what *Homo habilis* evolved from. If only there were paternity kits back in deep time which the scientists could refer to. Africa was the origin homeland of *Homo habilis*. Though even to make this statement will elicit controversy from some quarters. Out of Africa theories will be discussed below. Researchers never think as one unified body...kind of like our extended families.

Like all evolving species, we had cousins who shared some of our traits and abilities but perished in the hardship of the times, leaving little or no trace.

Over millennia, shifting climate changes moved people around the continents following the vast herds and edible vegetation. They had their objectives. Their goals were not to pioneer new territories or to set up outpost colonies (although in many cases that happened), but instead they were tracking plants and animals for foraging, scavenging and hunting. This is called the "first wave" of "Out of Africa". The waves of migration numbered at least three.

The Call to Move On

Contentedness was in short supply in the vast continent of Africa. Dissatisfied or displeased or maybe just extremely adventuresome, our early precursors started leaving Africa and populating other areas of the world. There were no roads and no conveniences, no maps, and no real idea where they were headed. Communication with those they left behind was impossible. Yet other lands and other shores called to them. Perhaps confidence, hunger or curiosity drove them on. Perhaps there was an innate notion of inquisitiveness that motivated archaic humans. Perhaps they ruminated over some sort of a declaration of independence and then got others in their group to go along with them. Maybe they lacked the one thing that makes things impossible to attempt: fear.

There were women, babies and little children among the wandering bands. Accommodation had to somehow be made for everyone. For the most part, these early members of our ancestors found most of the real estate was inhabited only by plants and game animals. It was a developer's dream. They could have made a fortune on the property market, but it was thousands of years too early. Timing is everything.

Granddad *Homo erectus*…. A Human in All but Name

The precursors to modern humans who left Africa were not *Homo sapiens*, the species that is us. Yet, they originated from *Homo,* the genus that includes us. From now on the term "modern humans" will be used to describe our species. Modern humans of several hundred thousand years ago are our tribe and our ancestors, we might as well own up to it.

The very earliest precursors were members of another archaic human species, *Homo erectus*. These were the first hominins to migrate out of Africa and adapt to a variety of different world environments. Though *Homo erectus* eventually died out, they remain the champions for long existence on the planet.

Homo erectus was a bigger-brained human than had ever previously walked the planet. They created and used double-bladed axes and came on the scene about a million and a half years ago. With more sophisticated tools and a need for more calories, new behaviors occurred in this species. As they moved out of Africa they brought their learned technology with them to many parts of the world. Their tools have been found widely distributed in Europe and Asia. They weren't as savvy as the techies of Silicon Valley, but their tools were better than using only teeth and hands for the tasks of hunting and butchering. And what good would a Smart Phone be at those tasks, anyway?

Eventually modern humans too, left Africa, and they met several other hominins in their travels: *Homo erectus* in Eurasia, Neanderthals in Europe, Denisovans in Siberia/Asia. These species were in existence and living at the time of early *Homo sapiens*. Property rules haven't changed: as time passes, the real estate market pressures build up with eager new buyers. Location, location, location.

Natural Selection/ When One Species Changes into Another Species

Let's tackle a subject that is often in the back of the non-scientist's mind: How does one species of the genus *Homo*, morph into another? It sounds like shape shifting. How, you might wonder, is it explicable that *Homo heidelbergensis* (who is often proposed as among the archaic humans that are in the genetic line to be an ancestor to modern humans) change from one species into another species...finally culminating in *Homo sapiens*?

Put quite simply, the explanation is that when one species changes into another species, the gene pool must change. To give an example, a species, probably driven by climate and a need for food, moves into a geographical area and over thousands or millions of years, being isolated geographically, gives birth to various offspring, some of whom are more successful at survival than others. Eventually the offspring that are best suited to succeed in the environment are the ones who endure. Through eons of time and thousands or millions of biological choices, their genes become genetically dominant. The newer models look and act much differently from that very archaic ancestor. All this takes place within vast periods of time.

Another theory which is an adjunct to the first, is that two different species meet, mate, have a few viable offspring, and again, over vast swatches of time and isolated conditions, the gene pool is changed. (See the chapter on Neanderthals and the chapter on Denisovans.)

This is the Darwinian model of evolution by means of natural selection that has long been upheld by the majority of the world's scientific communities. The theory goes farther than just human development over earlier periods of time. There is the ever-anticipated future. Our species, as suggested by scientists, will continue to evolve as a response to our living conditions - different food, geographical and environmental changes, the advent of modern technology and of course the amalgamation of our different ethnicities.

Species evolution and the forces that produce it have never stopped. Some people will always be favored genetically, and their offspring will be more likely to survive. It's impossible to say that we're evolving in any direction. Perhaps we'll end up for a while becoming a species with gigantic heads and frail, ineffective legs. Evolutionary pressures and modifications are always at work, adapting and changing us so we can deal with whatever circumstances arise.

For humans, the forces for change now and in the past are such things as new diseases, climate change, changes to the food supply, and new social and political selection processes by members of our worldwide tribe. When thinking of appearance, if we want the humans of the future to look differently, we might need to cut back on the Double Mac and fries.

Natural Selection

The process of gradual change of species is called *speciation*. Speciation is the descent of one or more new species from an ancestral species. It is also called "Natural Selection". This is a term that Darwin coined back 1859 in his book *On the Origin of Species*. The term "natural selection" started getting a bad reputation as far back as the 1940s because of the consequences of World War II. Many scientists now use one word, "selection" to indicate speciation because it is not a value-loaded term.

There will always be people who have opposing ideas to the theory of evolution, and its consequent natural selection theory, but the Darwinian model has been used to observe, explain and prove species adaptation and eventual evolutionary change in other animals with much shorter life-spans than *Homo sapiens*. If you have another explanation, other than one that involves miraculous intervention, you might want to write a paper about it. Or, if it is very simple, Twitter might be a good vehicle for spreading the word. It works well in politics.

Chapter 11: The Climate, the Land, the Animals, the Plants

The Ice Age of 2.5 million years ago which occurred when the genus *Homo* was a small and struggling group of proto-humans, was by no means the first glacial age in the history of our planet. Glacial episodes bring fierce, extreme and dynamic changes. The key to our resolute character was forged thousands of years ago by the forces that compelled survival for our adaptive genus against overwhelming odds in the African savannah. Over time, our personalities were molded by the severe environment in which we found ourselves. Our physical and behavioral adaptations were focused on surviving in the struggle with our greatest nemesis and most demanding teacher, the climate.

There was a refrigeration system in motion at the time of the arrival of *Homo sapiens* (us). The cooling was particularly severe in the far north where glaciers spread into Europe and North America. Much of the earth's water was locked in ice. The sea level dropped by over 300 feet. Surface waters of the ocean that cooled when glaciers expanded, yielded little moisture to the atmosphere in many regions. On the savannah of Africa, it not only became cooler, but the summers became progressively more arid.

<u>The World Before Humans</u>

Walking on two legs instead of four legs, our proto-human ancestors first arose in the forests. Three million years ago much of the land of Africa was filled with closely growing trees which meant that our two-legged cousins had alternatives; they could climb down from the trees or, if they chose, stay on the tops of them. Their bipedalism allowed them to move above the forest canopy by walking along the top branches on two legs while using their arms to support their weight, or if caught in an area with trees far apart, they could use their grasping hands and strong arms to swing to a nearby branch.

As time passed, shifts in climate led to dense forests being replaced by grassy plains with far fewer trees. The African climate dried out. Climatic conditions caused savannahs to appear where the thick forests had been. These sweeping grasslands established a new motivation for the early hominins. With trees spatially so far apart, it was best to permanently climb down out of them. The transition from tree dwellers into total ground dwellers happened over millions of years. It was the climate and the plants that were the motivators. The search for food now had to be negotiated over longer distances. Two feet and arms free to carry objects were now not just optional but imperative. At the same time, scavenging small animals from larger predator kills was becoming more popular.

The Introduction of Humans/ 2.5 Million Years Ago to 11,000 Years Ago

The forests had fragmented with only scattered trees and copses of trees. The grasslands became seasonally wet and dry as the ecosystem changed. The stage was set for the entrance of the genus *Homo*. Between the beginning of the last Ice Age 2.5 million years ago and its temporary retreat to a warming stage starting over 11,000 years ago, different struggling species of humans made their shivering debut and by necessity went through a heavy learning curve.

By the time some of the humans in Africa left their place of origin, human brains had stopped growing. Brains in *Homo sapiens* had now developed to the point of having gained more complex knowledge; ideas were becoming more advanced. Yet, the climate continued to dominate where humans would be able to populate new territories. Long term atmospheric conditions were the motivating force behind many human inventions, including their creations of stone, bone and antler tools for hunting, digging and scraping.

Despite the drying tendencies, there were fluctuations in the climate when it was less than bone-chilling cold. Nor was it always exceptionally dry. During these periodic warmer/wetter stages, when the climate was more like our present era, the human population grew into possibly the hundreds of thousands in regions such as central Africa, southern Europe, India and Southeast Asia. This was when the Earth's ice caps were smaller, the global sea levels were higher, and the equatorial regions remained warmer.

Yet, the general over-all drying continued during the time of human's arrival on the scene. This drying trend is evident in the mammals that evolved with us. For instance, grasses became harsher fodder; they evolved to contain tiny bits of silica that began to discourage archaic humans from eating it. The mammals that continued to graze on grasses, such as horses, adapted to the dry and tough grass by developing long, continuously emerging molar teeth with convoluted surfaces. This adaptive technique is still present in modern horses.

Our species had their own form of adjustment to the climate: we adapted by using our larger brains. This was admirable, but it also meant that the overwhelming waking hours of each day were spent searching for food to feed our active, calorie needy brains.

The Animals Among Us/ 60,000 Years Ago

As populous as modern humans had become during periods of warmer temperatures and wetter conditions, their population size was dwarfed by the numbers of animals that lived among them and which they hunted. Sixty thousand years ago the constantly changing global climate changed again and began to become cooler and drier once more. The forests continued to die out, and immense savannahs opened lands in Africa for incalculable groups of new animal species.

Flicka, a Meal, Not a Friend

The grassy expanses were home to vast herds of two different ungulates which became a staple for the humans' diets: reindeer and primitive horses. This might give pause to a more careful deliberation for those people who are considering a Paleo diet.

The World According to Herds

Beholding the scene from the advantage of the early humans, the animals on the savannah must have looked like the earth belonged to these teeming swarms. Today the profusion of animals in the Paleolithic period is difficult for us to imagine. Every current species native to Africa and Europe was there, along with many species that are now extinct. There were numerous swarms of birds blackening the sky in the daylight hours. Bats came out in abundance at night. There were rodents of many kinds that smaller predators were hunting. Modern humans were competing with the larger predators in pursuing small mammals. There was always a glut of fish in the rivers and lakes. Medium-sized predators like wolves, jackals, foxes and raccoons hunted in packs. Ferocious, large cats which were extinct forms of leopards and cheetahs were lurking in the bushes.

The cave lion, a now vanished species, was half again as big as today's lions and like our modern-day lions, hunted in prides. Like the humans, the cave lions were tracking and hunting aurochs (an extinct form of cattle), bison, giant deer, and elk. A magnificent creature called a megaloceros, now extinct, whose antlers were shaped like palm fronds and spanned twelve feet, was also hunted by both the cave lions and the humans.

It would be easy to forget that there were very small animals that were also on the hunt. Small creatures whose birth rate and population growth many times exceeded any of the mammals, were dogging not only the herds of animals but also our ancestors. If you spotted a group of modern humans 60,000 years ago, you would note that they were surrounded by clouds of flies, midges, gnats, and mosquitoes. All these insects were doing their own form of hunting.

Predator Animals That Were Vegetarian

The largest land animals at that time were the woolly mammoths. The herds of woolly mammoths were considerable, and their range covered the entire northern hemisphere from Europe, across to North America and Asia. They have been called "lumbering giants" for their size of six tons, but their food selection was so discerning that they could sort through different blades of grass, deciding on which ones they preferred to eat. Since their size was so massive, they often grazed close to twenty hours a day to consume the 400 pounds of food they needed.

Unfortunately for our own species, enormous cave bears also filled an ecological niche 60,000 years ago. Cave bears, a species which, like woolly mammoths, have since died out, were much larger than the largest of our modern-day grizzlies. Late in the fall, cave bears dug out large cavities in cave floors for hibernation. Woe be it to the humans who might have accidentally entered the cave of the bears. Cave bears were fierce, and humans surely were aware of their potential danger. Yet, powerful as their jaws were, cave bears were vegetarians.

What Types of Plants Attracted Us?

Wooly mammoths and cave bears were not to be reckoned with without careful consideration. Yet both these largest of the land animals were plant eaters. It is fair to ask what kinds of plants tempted humans 60,000 years ago? There is no doubt that large pieces of meat were devoured, but did any of our modern human ancestors pass the salad bowl? When reading about early humans, you might get the impression that back in the Paleolithic era, all men were ruthless hunters who supplemented meaty diets with the occasional fruit "chaser."

But the history of the human intestine proves this theory wrong. Our guts have evolved from fruit and leaf eating proto-humans. From the start of our species we have eaten a large range of fruits, nuts, and, yes, vegetables—especially fungus-covered tropical leaves. The vegetable matter was, and is, harder to digest than fruits. In today's world, vegetable material is what often gets called fiber. Vegetables are less digestible than fruits and to some of us, particularly the two-year-old who won't eat his broccoli, vegetables are less palatable.

Grazing on the Less Tasty /Vegetables

By the time of human's arrival, vegetables were growing all over the world. Early humans quickly found that vegetables are a species which contained some nutritious plants and some plants of no great nutritional importance which we have since dubbed "weeds".

Many of the veggies of long ago you won't find at a grocery store, like ferns and cattails. The archaic human's love of digging up tubers, particularly various forms of ancient turnips, is reflected in the thousands of stone digging tools found among their fossil remains.

The vegetable species which we eat today were consumed, but of course, in their non-hybridized form. Yet some vegetables like wild legumes have remained virtually unchanged and predate humans.

Modern kale, cabbage, broccoli, cauliflower, Brussels sprouts, and kohlrabi are all members of the same species, derived from a single prehistoric plant which was eaten by early peoples. Wild carrots grew in abundance, but until the Agricultural Revolution 10,000 years ago, when modern humans practicing agriculture intervened and developed a palatable thin tapering orange-colored root, the early peoples probably left the wild carrots for a Stone Age Bugs Bunny.

Proportion sizes of how much of early human's diets were vegetables is in dispute among researchers. The best guess is that human's intake of all nourishment differed significantly with location and opportunity. Many scientists think that although hunter-gatherers probably ate a far wider variety of foods than modern humans, they would have been cautious about consuming large quantities of anything they were unfamiliar with...wary of the risk of poisoning.

Some of the leafy greens and tuber plants probably were not too satisfyingly tasteful and took some getting used to. Assuming that survival is an innate driving force in all animal and plant species, vegetables and fruits differ significantly in their interest in encouraging animals (including humans) to eat them for the plant's reproduction purposes.

Because most vegetables have means other than animals to disperse their seed, they have no need to entice animals into eating them to continue their veggie existence. Consider the brassica of Europe and Asia: this genus contains cruciferous vegetables like cauliflower. These plants produce hard, small seeds that disperse through the air. It is the wind that supports its reproductive technique for planting new seeds in fallow fields.

The Human Love Affair with Fruit

Humans of 60,000 years ago definitely ate fruit. Small bits of date have been found stuck to the teeth of Neanderthals. Several of the fruits that we eat today have been around for many hundreds of thousands of years. There is a site in Northern Israel where archeologists have uncovered figs that date to 780,000 years ago. In the Paleolithic era there were olives, plums, oranges, pears and grapes which grew in forms that would be recognizable today. There is even evidence of ancient apple trees, growing many thousands of years ago. Perhaps there is something to that Garden of Eden story? None of the fruits were exactly like the modern crossbreed varieties that we consume today. But they would be both identifiable and edible.

The evolutionary purpose for plants, as with all living things, is to continue the species. Millions of years ago fruits developed sugary harvests which produced an encapsulated seed surrounded by a sugary fruit to encourage sweet-toothed mammals to gobble them and disseminate the seeds. Once eaten, digested, and excreted, the seed lies on the ground, free to grow, fertilized by its own small stack of manure.

Vegetables might very well not win the popularity taste-bud contest since they gain nothing from being eaten and so there wasn't the selection pressure to evolve sweet-tasting delectable roots and leaves.

This all means that since Stone Age humans lived in a world without chocolate, fruits were the next best thing.

A Modest Drunken Monkey Proposal

Ripening fruit may be at the bottom of an ancient secret: fruits may be the source of our love of alcohol. Ethanol, known better as alcohol, arises naturally in overripe fruit. Attraction to and consumption of ethanol by various primates may go back millions of years. Microscopic fungi, (yeasts) which live in fruit, get to work during the ripening process fermenting the sugars, changing the fruit to a plump, mushy cocktail. All primates including humans and many other mammals are attracted to the pleasant smell of ethanol. The faraway odors of the ethanol in the ripening fruit, drifting in the wind, would help them find scarce calories in tropical rain forests.

The fruits, however, have their own evolutionary objective: the continued generation of more fruit. *Australopithecus* and early hominins eating the alcohol- filled overripe fruit found that they not only could find food high up in the trees, but they could find highs high up there too. Or perhaps, the heavy ripened fruit was picked up after it had fallen off its tree. Looking at the situation from the purposeful aim of the fruit, its goal would be met if it enticed great hordes of primates and humans to eat it.

Further, alcohol stimulates the appetite. (If you ever drink aperitifs before dinner, you might already know that.) Not too surprisingly, the fruit washes its peels of any unintended consequences.

The drunken monkey theory, which might be more appropriately called the drunken human theory, is the hypothesis that eating over-ripe alcoholic fruit for hundreds of thousands of years has setup the human species to genetically crave the molecule of ethanol. If anyone you know has a tendency for alcohol abuse and attends AA meetings, they could blame it on fruit trees and the specialized human DNA when giving their testimonials. Hope this helps.

Chapter 12: Out of Africa/Pathways

There are many pieces of evidence that lead researchers to the conclusion that Africa was the place that *Homo erectus* and modern humans left. The researchers named this exodus, "Out of Africa", which is a significant clue as to which way the exit sign was pointing. Though there were many who stayed put, and for whatever reason chose not to leave, for many others it was definitely "out" that many *Homo erectus* and modern humans wanted. There isn't any indication that their migrations were connected to a forcible exit from another species. Instead, they were following the herds and the plants as the climate changed.

Both the *Homo sapiens* who left Africa and those who stayed in Africa were much larger brained than *Australopithecines* and apparently had no problems finding directions (though it must be remembered that about half the group were female). It sounds like the beginning of a bad bar joke..." On the way out of Africa my wife asked...."

There are three theories that are significantly different from one another concerning how our hominin ancestors evolved into modern humans (either in or outside of Africa) and began to wander all over the globe. They are called "dispersal theories". The distances traveled are substantial and remarkable, considering the journeys were not taken in an era of the combustion engine. Foot power is much more effective than we give it credit for.

<u>Theories of Passages Used to Journey into New Uncharted Territories</u>

<u>Out of Africa Theory</u> – suggests that *Homo erectus* evolved into *Homo sapiens* in Africa, and then ventured out of Africa and scattered to locations all around the world.

Multi-regional Evolutionary Theory – suggests that *Homo erectus* ventured out of Africa and then evolved into modern humans in several different locations throughout the world.

Southern Dispersal Route- suggests that an additional wave of colonization occurred between these two better-known dispersals.

There is evidence to support all these theories. In fact, it is possible that all three are correct. Some *Homo erectus* may have evolved in their homeland of Africa and eventually became the modern humans who left in large waves to begin to wander many places in the world. Also, some *Homo erectus* may have left Africa and then, through time, isolation, genetic evolutionary changes, and selection, *Homo erectus* may have developed into modern humans in various parts of the globe. And some waves of humans and/or *Homo erectus* may have left via the Southern route and dispersed and eventually displaced other hominin species as they ventured throughout the world.

These theories are all plausible. Evidence exists for each one.

Green Sahara Episodes

Leaving Africa was physically easier in some eras than in others. Many travelers out of Africa were assisted by what is called the Green Sahara episodes. These were occasional periods of time such as 250,000 years ago when the air was warmer and moister over northern Africa and the desert of the Sahara was green and vegetated. During these periods of warmer and more humid climate conditions, the Sahara contained a series of linked lakes, rivers, and inland deltas. These formed an inland waterway that helped to route both humans and animals northward. At that time herd animals and people could move more freely out of the African continent.

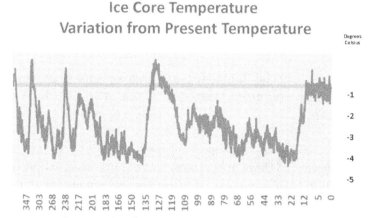

Around 250,000 years ago the earth's climate was in a warming period. Warming plus moisture caused the Sahara Desert to blossom with plants and vegetation that both the humans and the animals which they were hunting could use to their advantage.

Emotional Responses: Leaving Home Might Not Have Been Easy

What these humans felt when leaving their homeland is open to speculation. They were humans with large brains; even *Homo erectus* had a large brain, though not the size of a modern humans'. Interestingly, a hundred thousand years ago *Homo sapiens* had brains somewhat larger in size than ours today.

While getting ready to leave Africa, were they thinking of new opportunities? Did they pack anything as they made their long journey? Did they take souvenirs and reminders? Was it necessary to leave some of their relatives most dear to them? Did the children bring their favorite carved rocks, bone, antler or wooden play toys? Their expedition was more than a ramble; they would never again see the ones they were leaving behind. It is probably fair to think that they were aware of that fact.

Paths Out of Africa: The Southern Route/ Ethiopia and Beyond

For over a hundred thousand years small bands of modern humans were leaving their homeland and making expeditions out of Africa and into other areas. Exceptionally early fossils of what are essentially modern humans have been found in and out of Africa.

The southern route took people from Ethiopia, across the channel between Yemen on the Arabian Peninsula, and the Horn of Africa. It connects the Red Sea to the Gulf of Aden and beyond to the Indian subcontinent and East Asia. Ethiopia harbors some of the sites where there is evidence that the earliest anatomically modern humans stayed, or in other cases only encamped for a while. The fossilized skeletons from one Ethiopian site are almost 200,000 years old. In one location in Ethiopia, lying alongside one of the partial skeletons, blades and crude knife points have been found.

Also found in another site in Ethiopia are three skulls from humans that lived 160,000 years ago. Found with these skulls are Acheulean hand axes, cleavers, scrapers and blades [1]. Evidence from this wave of migrants from Africa is sometimes referred to as "Out of Africa 2".

Paths Out of Africa: The Northern Route Through Israel

The earliest known pathways out of Africa probably occurred following the route through the Levant, an area that today encompasses Israel. The name "Levant" means "where the sun rises" and describes the region's eastern location along the shores of the Mediterranean Sea.

Modern humans first emerged from Africa very early in their existence. It used to be thought that approximately 100,000 years ago a wave of modern humans left their African locations to settle in other places in the world. Though it is true that many modern humans left Africa at that time, recent discoveries have pushed that date back even further from 100,000 years ago to 300,000 years ago. These people are called "early modern humans" and their fossilized remains have been found in a cave located along the western slopes of Mount Carmel in Israel and nearby in a rock shelter called Qafzeh, located near Galilee.

Wandering seemed to be part of the nature of *Homo sapiens* whose thousands of years old remains would eventually be found in regions of the world as far apart as Tasmania and Chile.

The Distance Traveled

In case you think that those distances seem too remote for people to travel on foot, below is a mathematical account of how distances add up as the years go by:

If every 15 years early modern humans had a new generation and at least part of the new generation moved 5 miles from their family home, then:

After 100 years the latest generation would be 35 miles from their ancestral starting point;
After 1,000 years the latest generation would be 350 miles from their ancestral starting point;
After 10,000 years the latest generation would be 3,500 miles from their ancestral starting point;
That's about the distance from Olduvai Gorge to Israel.

Multiple Out-of-Africa Migrations

This doesn't mean that large waves of human migrations were constantly leaving Africa, but rather that small bands were leaving their homeland earlier than 100,000 years ago. Some scholars have pointed out that *Homo sapiens* followed some of the same routes out of Africa and into Eurasia as *Homo erectus*. It seems once a good route pops up, people stick with it.

Pushing Back the Timeline: Misliya

A jawbone found by paleoanthropologists in a collapsed cave (called Misliya) in Israel has been shown to be between 194,000 and 177,000 years old. Blades and other stone tools were also found in the Misliya Cave on Mount Carmel's west slopes, near Haifa. The humans who lived there were making a comfortable place for themselves. Among the discoveries are several hearths. Near the hearths was plant material that researchers believe was used as bedding. The group who encamped at Misliya appeared to be capable hunters of large game animals. Charred bones of big animals like deer and gazelles were found in the hearths which indicate that fire was controlled and used for practical purposes. The Misliya people were probably nomadic, following prey animals or leaving the area when the winters got too cold. This new fossil find makes the modern human exit from Africa much earlier than scientists realized.

Is That Uncle Flint? /Or Maybe Not

There is some reason for caution about deciding how closely related these people were to modern humans; the ancient people from the Misliya Cave shared some bodily characteristics with modern humans, but not all. If scientists could revive the body from this oldest-of-old human being, and you met the almost 200,000-year-old in a dark parking lot, you might not be overjoyed about coming nose to nose with a relative who existed that far out of the past. (A dedicated paleoanthropologist might be more thrilled than an average person; scientists are sometimes incautiously devoted to their field studies.)

Researchers studying the earliest human skeletons have suggested that "morphologically" (meaning how the body would have looked) the skeleton found in the Misliya Cave was "more closely related to us than to Neanderthals". That would be faint comfort if the guy had a double-bladed axe in his hand.

Nor does finding these early human skeletons necessarily mean that one of them is your long-lost Uncle Flint. It looks as if early humans made many short-lived excursions all the way to Eurasia thousands of years before humans finally mastered successful travel and permanent settlements. Current thinking is that this group from the Misliya Cave might not have lived to pass on their genetic code.

In the case of the Misliya Cave and the fossils found in that location, there is a possibility that Uncle Flint belonged to one of those mysterious societies of *Homo sapiens* who were too busy with their fraternal organization (perhaps called the Benevolent and Protective Order of Wooly Mammoths), to have had much time for other things. The humans who lived and expired on that rocky outcropping in a cave in Israel probably died off without leaving any trace DNA behind. Bands of brotherhood can be taken too far.

Scientists who have been studying the internal and external shape of the skulls found in both Ethiopia and Israel have discovered that based on their physical characteristics there is a link between the early humans in those two locations. So, the ancient hunters in the Misliya Cave had family connections. We're relieved.

Jebel Irhoud in Morocco: Oldest Human Fossils

The members of *Homo sapiens* who stayed behind in Northern Africa had been living there generations longer than scientists originally estimated. There is an archaeological site in Morocco with relatively modern looking human fossils that are 300,000 years old. These are among the oldest fossil human remains in the world. This is not to say that researchers might not find even older human skeletons in the future. One Israeli scientist is quoted as saying that he thinks humans may have existed for 500,000 years. So far there is no fossil evidence to prove this very early date. But there are many sites that have yet to be found;

surely there are ancient spots that have been unknowingly passed over for millennia. As one researcher put it, "absence of evidence is not evidence of absence".

At Jebel Irhoud there are five skeletal remains of early modern humans. These humans were busy hunting and butchering the prey that they were stalking. Many carved pointed stones and scrapers were found with these skeletons. The tools date back 350,000 to 280,000 years ago, in some cases dating even farther back in time than the human fossil remains. The fossilized bones of the animals they were eating show evidence of being cooked al dente, and then when done, the flesh was scraped off the bone and eaten along with the bone marrow.

Speaking Relatively/The Humans Are Modern

In structure, the skulls found at Jebel Irhoud were not entirely like a modern human. They had the archaic features of an elongated and low braincase. Yet, what if they weren't going to pass an advanced test in algebra? This is deep history; they weren't going to apply for college admission anyway.

Paleoarcheologists have pronounced the skeletal remains at Jebel Irhoud to be "modern-looking humans". But this is a relative term meaning, in this case, that these people were not Neanderthals and were not *Homo erectus*. It is worrying to think that the fossilized skeletons found at Jebel Irhoud are humans still looking for a name. Perhaps they need a good PR person to update their identity.

The Humans Were as Varied as Heinz 57

There is no clear dividing line between late *Homo erectus* and early *Homo sapiens*. Many fossils between 500,000 and 200,000 years ago are difficult to classify as one or the other. Though at present it is believed that the speciation of *Homo sapien* divergence was around 400,000 years ago, there is a plethora of human fossil remains that have been found recently where it is unclear what category they should be ascribed to. What seems very possible is that migrating modern humans interbred with local regional varieties of archaic humans. In some cases, the DNA was carried on and in other cases it was not.

After coming out of Africa, the amount of interaction *Homo sapiens* had with other types of hominins during their sojourns is still widely debated. An example of a mixture of *Homo sapiens* possibly mating with other "hybrid" species is a jaw bone fossil found in Taiwan. Called *Homo tsaichangensis,* scientists think it is over 100,000 years old. At this time researchers have ascribed the bone to be a *Homo erectus*, but many think it is a hybrid species, a cross between *Homo erectus* and an unknown roaming group of humans. Since the jaw bone was found by a fisherman in the waters off the coast of Taiwan, we can get a picture of what early wandering humans, close to a romantic coastline, might have been up to.

One way or another, *Homo sapiens* are descendants from migrating groups plodding their pathway out of Africa. There may have been many times when the rough folks of the country looked quite attractive to the traveling, more modern humans. The archaic (and in some cases hybrid) humans might have been provincial, but their beds were warm, and they made a great rhinoceros soup. Unless you've been on foot trips over vast lands where all you had to eat is what you foraged and hunted, don't judge.

(1) Acheulean tools like handaxes and cleavers were made by Homo erectus. Usually made from flint or chert, they are far more complex in their construction than the earlier stone tools made by Homo habilis.

Chapter 13: Transitioning into Full-Time Human Behavior

The desire to expand and leave the Mother Land of Africa didn't start with modern humans. Starting 1.5 million years ago, *Homo erectus* had already settled in Asia. *Homo erectus* settlements are found throughout most of Eurasia, some of the settlements look very much as if they were planned with huts surrounding a large fire pit and small hearths. Neanderthals, too, left; in their case their homeland was northern Africa. They were much later in time; it was 300,000 years ago when Neanderthals settled in Europe and Asia. The exact time of arrival of modern humans into Europe is uncertain. By 70,000 years ago humans who looked like us today, were traveling out of Africa.

Home Is Where the Heart Is

Homo sapiens came from Africa, a continent that was not touched by ice sheets. Near the equator, where many of the *Homo sapiens* resided, much of the time the weather was more tropical. For our ancestors, traveling through the cold climes of Northern Europe to get to Asia was a slog; residing in chilly Northern Europe was problematical. Compared to the tropics, the farther north in Europe they traveled, the closer it is to the poles. The length of day varies; in the winter, days become very short, dark, and severe in comparison to the warmer, drier equatorial climes.

Growing periods of edible vegetation in northern Europe were short. Gathering food supplies would have been reduced. Only the most skilled hunters were going to survive the northern winters when the plant resources would have been unreliable. The humans needed to use all their knowledge and abilities. Expertise about how to control fire would have been essential. Scavenging for dead animals in case hunts were negligible would still have been a useful skill. Who really wanted to set up shop in nippy northern Europe?

1.5 Million Years Ago/*Homo erectus*

By about 1.5 million years ago, *Homo erectus,* who had already started migrating out of Africa via the Southern Route, was beginning to enter Southern Europe. The most ancient route into Europe was via the Levant – a corridor connecting northeast Africa and western Asia where countries like Israel and Lebanon are today. Over a million years ago, archaic humans settled first in warmer southern Europe; places that are now France, Italy and Spain.

Modern Humans Push on to Europe

As early as 300,000 years ago small bands of almost modern humans (us), tried out forays into the Middle East. Though their campsites, fossilized skeletons, and a few stone tools have been found, these small bands of early humans did not live long enough to establish any kind of colonization. Their settlement, if that is what it was, wasn't successful. A hundred thousand years ago modern human groups again tried their hand at moving out of Africa and into Europe with very minor success.

Success Happens After 50,000 Years Ago

The achievement of semi-permanent and seasonal settlements of modern humans into Europe started sometime after 50,000 years ago. By this time our species, *Homo sapiens*, were migrating out of Africa in vast numbers. This is called "dispersal" but that is too benign a term for what happened when humans traveled to other lands and permanently applied their hunter-gatherer technology to the new land around them.

The great success of the human diaspora into lands outside of Africa was a gigantic achievement, an accomplishment that paleoanthropologists have ascribed to various auspicious occurrences. The theories range from a bigger brain size, to new and better stone equipment and weapons, to new knowledge and collective language, to climate change.

As modern humans dispersed, they began encroaching on other hominin and animal species' territories. This was a slow process, but modern humans gained the upper hand, eventually leaving only the memory of many animal groups and leaving no members of any other hominin species to tell the tale.

Only Us / Dominion by Humans

There is now only one type of humankind left on earth: us. Whatever the cause, and there could be several overlapping causes, Neanderthals and other archaic hominins in various places both within and outside Africa eventually became as extinct as the dinosaurs. The question is how did *Homo sapiens* become the species who are not only the one humans who are left, but how did they ultimately manage to exert dominion over all the other creatures?

The Role of Cooperation

A plausible and popular explanation among paleoanthropologists is that we are team players. If humans are gathered in large groups, eating and sleeping together, and most importantly, sharing the same goals, then people must be cooperating. If, further, this process takes generation after generation to accomplish, cooperation must be passed down to all future descendants. Our ability to cooperate, even with the people who are not blood related to us, is one of the hallmarks of our species.

Cooperation among same species is not unique to *Homo sapiens*, think of various species of bees, ants, termites or naked mole rats. But mutual collaborative goals and agreed upon cooperation is realized in the extreme among humans. Since we are relatively defenseless on an individual level, the ability to unite and work together is a key to the survival of our species.

We Take This Moment to Remember the Dearly Departed Neanderthals

Homo neanderthalensis (400,000 - 30,000 years ago) might have had their own idea about what it was like to live among modern humans. But they didn't live long enough to get facetime on social media. In their final decline, the last Neanderthals seem to have disappeared in small and isolated cul-de-sacs. All that is left of this archaic species is a miniscule remembrance of them evidenced in our DNA.

Departed Members of Denisovans, Extinct Because of Us?

Though not much is known about the Denisovans whose remains have been found in a cave in Siberia, it is known that more than a chance "passing like two ships in the night" had taken place between our species and theirs. Some DNA evidence survives in living people in various regions of the world. The Denisovans must have been on the move at one time. Evidence of their existence in Siberia and DNA evidence from New Guinea and Australia shows that distances were traversed, possibly not only by *Homo sapiens* carrying their DNA, but also by the Denisovans. Yet, Denisovans have also gone extinct.

A Story Book Tale Without A Storybook Ending

There are stories about how, through time, the meek were vanquished and the less than noble succeeded. It is an old but persistent story that at one time there was a lost world filled with peaceful, egalitarian hunter-gatherers who were replaced as time went on by destructive, death-and-property obsessed moderns. There is partial evidence to support that story. Everywhere modern humans settled, archaic humans went extinct. This is also true for a vast number of animal species. The question remains as to how peaceful and egalitarian the previous residents of that lost world were.

50,000 Years Ago, And Change Was Still on The Horizon

After spreading out from the Horn of Africa, across to Asia, Indonesia, and Australia, some of the *Homo sapiens* who anatomically resembled us decided to make a home in Europe. Though Europe was still in the grip of the Ice Age, these hunter-gatherers were able to successfully inhabit these cold climes. They had the courage and necessary skills to leave Africa and venture into new territories, yet for the first several thousand years of their settling in Europe, their activities were not much different from any of their ancestors such as *Homo erectus*. The early modern humans settlers had made a small inroad in developing better tools, and they could effectively hunt for a wider variety of game than their archaic cousins the Neanderthals. Even on rare occasions they created decorative objects or grooved a pattern of lines on a rock or shells. But they were a group of *Homo sapiens* that erred on the side of utility, not originality.

43,000 Years Ago, /An Explosion of Imagination /Cro-Magnons

Then, about 43,000 years ago the ancestors of the people who had left Africa and settled in many European sites, made some large-scale changes. These were the same modern humans who had been living in Europe for several thousand years. *Cro-Magnon* is the name given to these humans, a name that comes from the location in France where their remains and relics were first found. "Cro" is a French dialect name for "hole" or "cavity" and "Magnon" is the name of the owner of the land where their remains were first found. Cro-Magnons are essentially Paleolithic Europeans, anatomically the same as us.

There must have been something in the water because suddenly Cro-Magnon people became the inventors, the creators, and the all-around clever people who were about to run the show. (Pass the bottle of Perrier, please.)

Exactly what cognition transformation triggered the new, developmental changes in these humans who settled in Europe, no one yet can say. But Cro-Magnons started to exhibit traits that can only be called revolutionary. These *Homo sapiens*, who looked like us, began to behave much like we do today. As the Cro-Magnons established themselves, they started to produce never-before-thought-of carvings, figurines, art and weaponry.

The cognitive revolution had begun. It was a time of creativity and new fabrications. Inventions like the first bone needles with eyes are found in this period. Creating clothing by stitching hides together to fashion fur garments made it possible to bear the cold. Cro-Magnons also stitched shells, beads, and animal teeth onto their cold-weather clothes. Not only did they start sewing their clothes, but they used string to fashion fishing nets and carrying bags.

Cro-Magnons also began to perform elaborate burials where they dressed their dead in clothes and decorated the dead with necklaces made of animal teeth. Red ochre was smeared onto the bodies of their dead. There is evidence that the living also decorated their bodies with ochre, perhaps on ceremonial occasions.

Trade and Cro-Magnons

The sea shells which show up in Cro-Magnon campsites and in their gravesites had been transported, along with other raw materials, vast distances. It is likely that rudimentary trade was taking place. Humans were beginning to develop more complexity in their attitudes and habits of mind. The emergence of trade, even if only in exchange of sea shells for fur hides, marks a very different outlook. Trade implies that certain items are valuable outside their intrinsic nature. This is a cognitive development that will eventually have revolutionary effects. One of the benefits of trade is that of building workable relationships with others based on the demand for goods they have acquired or created.

All these activities indicate that Cro-Magnons had a much richer culture than any of the earlier hominins. An explosion of imagination, inventions, and ideas had begun; nothing like this had been present before.

Weapons for the Hunt

Cro-Magnons brought with them the technology their ancestors had learned in Africa. But their talent for creative thinking was such that once they were in Europe they were making strides to change, create and improve.

Their weapons began to change as their technology advanced. Like the Neanderthals, the Cro-Magnon population was a hunter-gatherer culture. Also, like Neanderthals, they hunted big game like mammoth, horses and reindeer. But their tool-making ability quickly soared far beyond that of the Neanderthals. Elegant stone projectile points found at Cro-Magnon sites reveal that they became sophisticated big-game hunters. They also fashioned artifacts not only from stone but also from bone, antler, and the ivory of mammoth tusks.

The Handaxe For Hunting

Though hunter-gatherers had developed the handaxe over a million and a half years before Cro-Magnon, for obvious reasons the handaxe was never a useful tool as a weapon for killing big game animals. A handaxe can't be flung at a large animal with any kind of effectiveness; to get close enough to kill a large animal with a handaxe was extraordinarily dangerous.

The Evolution of Projectile Weaponry/Where It All Began

In the history of hominins, the development of an efficient weapon for killing wild animals developed from crude tools. In time, sharp stabbing sticks became the preferred weapon. This evolution took several million years to grow and improve. Wooden spears with honed points were used as early as 500,000 years ago.

Eventually advancement was made from stabbing sticks of simple wood with shaved points, to binding the stick with a sharpened stone. Researchers have found triangular stone-tips that are hundreds of thousands of years old that have wear markings showing both the tell-tale damage of a stone tip that is used for spearing and the hafting marks where it was connected to its wooden shaft. This required that a stone point was either glued or tied to a slender wooden or bone pole. The advanced technology became the weapon of choice as hunters went after larger game.

Cro-Magnons were now making refinements from their flaked tools; they created wafer-thin projectile stone points that could be mounted into bone or wood. These new inventions produced weapons that were capable of being launched at high speeds and long distances. This was the start of what is today an unbelievable growth of munitions; human resourcefulness and creativeness in the development of weapons had only begun.

<u>The Amazing Atlatl/See It Fly Through the Air</u>

About forty thousand years ago a remarkable invention of weaponry was invented: the atlatl or spear-thrower. An atlatl is a notched stick that a tipped spear is placed on so that the lever arm of the thrower is extended. This device allows the hunter to stand significantly back from the game animal and use the leverage of his throwing arm to achieve a great velocity when propelling his spear. The atlatl allows the projectile spear to fly at speeds over 90 miles an hour. This new weapon revolutionized the hunting of large game animals. Cro-Magnon hunters could cooperate in hunting parties, each hunter armed with an atlatl helping to track and kill perhaps several large game animals in a day's hunt. Butchering might have been done by the entire tribe and large chunks of meat carried off to the camp by the stronger women and young boys.

The New Inventions

The original cause of the human advancement in invention and creativity for hunting is not fully known. One of the theories for the rapid and ingenious development in arms is that different human groups were in competition with one another for hunting and mating purposes. Another theory is that modern humans were increasingly social and interacted with many various groups around them. This socialization and interaction would allow people to share concepts and ideas with one another. These are theories, and like many others, are not mutually exclusive.

Transmission and Social Behavior Through Language/40,000 Years Ago

The combination of the forces of changing climate, and outside competition is not enough to explain our successful worldwide evolution and how we ended up the only humans left on the planet.

Forty thousand years ago we see modern humans exhibiting new complex social and cultural behavior. Some of this behavior must have been the effect of genes passed on from one generation to the next. But humans had special skills that set them apart from all other species.

Transmission of knowledge was beginning to have a cumulative effect on the creative and social behavior of our ancestors 40,000 years ago. Ever since the use of human language, and its concomitant accumulation of knowledge through the generations, the human species has been exhibiting greater and greater evolutionary advancement. Communications through language made it possible for parents to tell their offspring what they had learned from the previous generations. There were probably many methods that were used for oral transfer of information including teaching, storytelling and apprenticeship.

Members of the tribe no longer had to learn everything through experience. There had been a leap in social complexity due to enrichment of information. For instance, each generation no longer needed to reinvent how to create the stone tools of their ancestors. This freed them to think about designing new and better implements. Their new independence made it possible to invent and advance fresh ideas. Forty thousand years ago these new approaches were reflected in many ways, including not only improvements in weapons but also creative and original art works.

Complexity Brings with It A New Set of Problems

The modern humans with superior weapons had brains capable of complex emotions, sentiments that went beyond the primal urges of hunting, sex, and aggression. Awe, wonder and imagination are probably not recent evolutionary inheritances. As complexity continued and groups became inheritors of generations of memories, there would have been new outlooks and complications. Just as now, within everyone in the past there must have been feelings of trust, love, hatred, suspicion, admiration, envy and various levels of tolerance and sociability. These would have varied in intensity depending on the person. All these emotions are with us today and are the stuff novels and movies are made of.

There might also have been old grudges. The grudges could have been the Stone Age version of what drove the murderous division between the Capulets and the Montagues. Who knows what went on in those primitive huts and on cave ledges? Though the time is thousands of years ago, anatomically modern humans in the Stone Age might very well have been capable of passions like the ones we have today. With brains a bit larger than ours, Cro-Magnons probably had the capacity to plan, deliberate, contemplate and sometimes stew over old resentments.

The Quest for Meaning Among Our Modern Human Ancestors

Forty thousand years ago it is likely there were some members among the human tribes who pondered, at least in some way, the human condition. The quest to appreciate the level of understanding of reverence, beliefs, tribal loyalty, sociability or antipathy and what is perhaps instinctual versus what is learned behavior, will be the subject of our next chapter.

Chapter 14: Tribalism, Rituals, and Beliefs

The humans who lived so long ago had no way of knowing that we moderns might be interested in their rituals and belief systems. It would be nice to think that if they had known, they would have left more evidence for us to mull over. Since there are only a few signs that we can draw upon, a lot of the ideas about an inchoate Stone Age religion or any form of belief, must for the most part, be conjectural. Thankfully however, there are some clues.

By 50,000 years ago, modern humans were performing activities that seemed to have nothing to do with hunting and gathering. Red ochre was being used on their bodies, not only on the bodies of the dead, but also on the living. By 40,000 years ago, humans as well as Neanderthals were burying their dead, sometimes with elaborate decoration.

What is Meant by Religion 40,000 Years Ago

To talk about religion in an age of stone, bone and antlers, when humans were smearing their bodies with ochre seems as if the implausibility of the topic would overwhelm the subject. Anything approaching an institutionalized system of religion which is upheld by a structured establishment is inconceivable in this very different time. Yet, there were humans 40,000 years ago who seemed to pursue certain activities with great devotion, such as the practice of burying the dead, which gives some possible indication as to an explanation about Stone Age rituals and beliefs.

Definition of Religion

Religion is arguably one of the most defining cultural inventions of humans. Certainly, religion is a powerful catalyst for human behaviors. For the Stone Age person, their beliefs might have served the same functions as they do in contemporary times. As described here, religion will be defined as a belief that brings an explanation for why things happen, gives a conviction that there is a purpose, frequently provides a sense of some superhuman controlling power that might respond to human entreaties, and provides a structure for group (tribal) attachment.

There are many theories about how and when religion came about. Since humans are the only ones on earth who practice religion, a great deal of interest is directed to its genesis. The two major theories which will be discussed here are the *concept of personal mortality* and the *theory of complying with group conventions*. These two theories weave religion in and out of their tapestry of primal human motivations. If you are curious and ambitious, you can find enough books about the beginnings of belief to occupy you for the rest of your life.

The Concept of Personal Mortality: Mortality and the Birth of Religion

There is a possibility that religion grew from a sense of countering the feeling of insignificance. The death of people, particularly the death of family and friends, creates an uncomfortable cognitive state, one that might linger for quite a while in the minds of the larger brained humans of 40,000 years ago. The presence of dead bodies could create a realization that death comes to everyone. Also, since death appears uncontrollable, the individual might be overcome with a feeling of inconsequentiality. To counter this feeling, there may have been steps along the path that helped people in their search for meaning and significance and was the origin of early forms of religion.

Altruism & Grave Goods/A Step Toward Religion

Intentional burial of grave goods along with the dead body may be one of the earliest detectable forms of altruism. Perhaps the grave goods were a symbol of reassurance to the dead. Grave goods would reinforce the idea that there is an afterlife, which might comfort the living. Outfitting the dead for their journey to the afterlife would create a sense that the living had helped the dead to go in peace to a place that transcends daily life. Selecting the kind of grave goods would be choices that were carefully made, creating the behavior of altruism. Sometimes the grave goods might have been items that the living were sacrificing by placing them in the grave, out of reach forever. The items that were sacrificed and purposefully placed alongside the dead might have anchored the archaic people to respect a form of moral code: what is the honorable thing to do?

Neanderthals appear to be the first humans to intentionally bury their dead. The early graves were generally shallow, and most of the grave goods were stone tools and animal bones. The animal bones may have indicated a type of food offering. By 40,000 years ago modern humans were also performing burials of their dead and including grave goods.

A Possible Basis for Ancestor Worship

It is thought that both the Neanderthal society and the human societies started out by burying their family members. Grave goods were an item or items that were considered valuable but were forfeited for the sake of the deceased. By burying their dead kin, the standard might have been set for venerating these dead loved ones. Eventually this form of sacrificing grave goods and honoring the dead family member might have been the basis for ancestor worship.

The Narrative of Ever-Lasting Life

A possible balm to the uneasy feeling of the inevitability of death might be to not only bury bodies with honorary grave goods, but to make up a narrative for the dead of a literal immortality, a promise that life never dies but continues in the afterlife.

Acceptance in a life after death could be considered an early form of belief on the path toward a religion. This type of trust in an afterlife has been explained by some scientists as an evolutionary adaptation to stimuli that is inexplicable and for which there is little human control.

The researchers further argue that this evolutionary adaptation is advantageous, citing many studies that show that there is a positive relationship between religious involvement and good health. In this scenario, natural selection would have favored those hunter-gatherers who practiced good health activities like honoring their dead and believed stories about a cosmic significance for the eternal dead.

Since the Dead Are Not Actually Dead

The leap from a self that is now believed to be meaningful after death to a relevant living self is not too far a step away. And if the hunter-gatherers had performed the death ritual "correctly", there might be a reason to believe that the dead would still be around and invisibly looking out for their still-living benefactors.

This might have led to some form of animism: the belief that things in nature like trees and sky have some supernatural invisible force that animates them. The reasoning might be, if the dead can come back to life, why not other things in nature? Rock falls, shooting stars, and lightning strikes, might be thought to have been generated by some unexplainable animating force. Dreams, visions and stories could all add to the mix. At 40,000 years ago, all humans could speak what paleoanthropologists theorize would be a proto-language. What better entertainment than to sit around the blazing fire and exchange ideas about the supernatural?

Anthropomorphizing the Deities

At the core of most religions is a personified deity. Eventually, the idea might come to our hunter-gatherer relatives that there is a god or are gods somewhere that are motivating natural events. How far back the instinct goes to humanize God is not known. The cognitive ability to think of creating a god in our own image was certainly possible at this time, but the record is probably secreted in the distant, circulating winds that first heard the stories of our long-ago ancestors.

In time there would be thoughts of connecting with the unknown god or gods being worshiped. Chanting, praying, singing or otherwise giving respectful acknowledgement to the gods in some form of ritual, would open a possible opportunity to communicate with them. In this way, some researchers believe, is where the gods are invested with super-powerful, super-knowing capabilities.

Through the power of ritual worship, the god(s), with its super-human capabilities, might be believed to be accessible. It has been speculated that there might not have been much difference between what we today ask help with and what the early humans might have been asking of their Paleo Gods. Answers to the challenges of life haven't changed: birth, understanding life, relationships, health, sex, hunger, gains, losses and death. The amount of negotiation skills the Stone Age ancestors had with their deities has not been quantified.

Demons and Devils

When people believe in invisible forces, whether it is in the form of kinship worship, animism, or belief in god(s), there is always a possibility of belief in a darker, more menacing side. There might have been some hunter-gatherers who believed in imaginary predators or in demonic possession. These beliefs could have followed from ideas of what we would now define as ghosts or spirits.

The Theory of Complying with Group Conventions: *The Social Identity Theory*

Humans seem to have an intrinsic trait that some scientist say is hard-wired; it is the instinctual urge to belong. There are recent anthropological studies that show that human's loyalties, opinions, and religions are shaped by the need to fit into a group. Possibly due to their primal need to cooperate within a tribal unit, humans are social animals. We share this trait with other primates. There is a reason great apes huddle in troops; they are born into a group and all their lives they stay in a collective unit and socialize.

In the time of deep history, humans were not that much different from their distant primate relatives. For early humans there was no such thing as a rugged individualist who struck out on their own without kin or group support. Or if there were such humans, they have not passed on their Ayn Rand genes to future generations. We stay in groups (tribes) and our social life is based to some degree on fitting in and helping within our group.

The "social identity" theory suggests that our concept of who we are and what we believe is shaped by the groups we identify with and people we live among. Successful survival in collectivist hunter-gatherer communities requires a conformist, homogeneous population. The safety of belonging to a group brings a wide range of advantages. Some of the positive sides to conforming and belonging to a group are: support and assistance for hunting and gathering; aid in giving birth and raising children; cooperation in cooking; help with constructing shelters; shared skills in creating weapons and other knowledge; and protection against predators and enemies.

But the price for membership in the collective is conformity and obedience to the band or tribe. The cost for fitting into the tribe means an assigned role for everyone within the group. Getting along with others in the tribe and following certain accepted norms are part of the bargain. A standard for what is an acceptable belief and what to do in different situations can be regulated by the group. Any moral imperatives in a collectivist hunter-gatherer culture are probably confined to social roles and duties to the tribe.

Gossip Used as A Method to Keep Standards

There are many ways a tribe keeps control of its members. Moral imperatives are not carried out by the cop on the beat. There is not an official peacekeeper in a hunter-gatherer group. Yet, rules need to be implicitly maintained. Together, the hunters decide what prey animals to follow. Together, the coupling of certain males to certain females needs to be upheld. This is true today even in polygamous tribes.

Anthropologists studying the few remaining hunter-gatherer tribes have found that about two-thirds of everyday conversation is gossip, with the clear majority of it being negative. As soon as some form of language existed, it is possible that gossip was a controlling mechanism for group conformity, even thousands of years ago. One of the attributes of humans is that they are interested in the affairs of other humans. Gossip is an effective line of communication, informing at the same time it regulates.

Once a standard mode of behavior is breached by an individual member of the tribe, gossip becomes a powerful technique for shaming. How far back in human history the element of shame goes will never be known. If members of a tribe are connected by trying to maintain a tribal standard of behavior, shame is a way to enforce group conformity. This is part of the "social identity" theory. Forty thousand years ago, gossip with the goal of shaming might well have been a weapon of the collective; many tribal members holding onto norms of behavior against an outlier.

The Dark Side of Conformity/The Primacy of Tribalism

Obedience is closely intertwined with conformity. Conformity can be as simple as everyone in a tribal group agreeing that a shake of the head one way means no and a shake of the head the other way means yes. Heeding the tribe's general procedures is necessary for the advantages of the tribe to continue. Whether it is following the behaviors for successful hunting and gathering or conforming to the language of the tribe, obedience to the group's fundamental long-held conduct is not only necessary, but critical.

Conforming to group behavior can have comforting effects. But conformity can also have dark consequences if the need to conform, belong, and obey the group passes all reason. In experiments on contemporary humans, scientists have observed that people, without any outside trigger for tribal behavior, will often form in-group loyalties. These loyalties frequently will make the human subjects guard against any contradictory evidence, choosing to indiscriminately agree with the majority in their group.

Some results of conforming to negative and destructive group norms are bullying, shunning, expelling or killing. These loyalties to in-group behavior may well have roots in the long-ago psyche of 40,000 years ago.

<u>Examples of Rituals of Conformity Having No Favorable Universal Attributes</u>

The pull of conformity and obedience may bring out group allegiances that focus on destructive elements. There is always a possibility that many could be led to behaviors that have no redeeming virtues and are not sustainable.

A Neanderthal burial discovered in Teshik-Tash in south-east Uzbekistan was of a 12-year-old boy whose body had been "defleshed" perhaps as a ritual preparation. Goat horns were placed near the body, and perhaps had some ritualistic meaning.

The body was buried close to the tribe's hearth. Could there have been a feast? It is difficult not to speculate.

Whether this is maladaptive conduct to group behavior or some special religious ritual carried out by only a few members of the tribe is not known. There are indigenous hunter-gatherer tribes who, up until modern times, practiced endocannibalism. This is the ritual of eating one's dead relatives to return their "life force" to the group. In this ritual, the dead person's power is spread around to the living by means of eating them. Whether or not endocannibalism was common throughout the Stone Age is unknown and controversial. No one really wants to think of their archaic relatives having had a barbeque with their family's dead bodies.

Bones from Neanderthal bodies were found at Gran Dolina, a site in northern Spain that dates back 800,000 years. They are the skeletons from several Neanderthal adults and children. The children's bones are covered with marks which suggest they were cut with stone tools. While the damage could be due to burial practices, most researchers who have examined the fossil bones think that the Neanderthals were victims of ancient tribal cannibalism and belief customs.

Possible Indication of Hunger

There are indications in some ancient sites that relate to possible famine. There are several times in specific areas where Neanderthals and/or modern human tribes were on the brink starvation. When people are extraordinarily hungry, they are less charitable and less empathic. Highly distressed people tend to prioritize their own needs. During starvation, bodily proteins start to break down. Essentially the body starts to cannibalize itself. In such extreme conditions, perhaps there were times when humans decided to cannibalize their fellow tribe member first.

We Galvanize Around Leaders

Forty thousand years ago, what would be the effect, listening to stories about the supernatural? It can only be conjectured, but a guess is that the narratives might have an enormously persuasive influence on insular, tribal peoples. There was no science; knowledge was primitive. Even communication with outsider tribes was probably limited to questions about hunting, locations of edible plants, danger spots of the land, and discussions about children.

It is possible that stories were limited and few. Perhaps one narrative about the mysteries of life prevailed and there were no contradictory stories. The Christian Bible might have it right, in the beginning there was the word, or to be a little more precise, the narrative. The tribe might relate to a tale from a leader who could recount a forceful, convincing story. Stories can have hypnotic effects when delivered by a strong leader. Tales of how there might be a "real" world inhabited by the invisible-unknown might resonate with urgency in the minds of members of the tribe who were searching for answers.

Ceremonies and Rites

Ceremonies would also be persuasive. Burial of the dead with ceremonial rites could have moved the emotions of tribal members. Only humans, among the many animals in the animal kingdom, are emotionally stirred by words, ceremonies, and rituals. In the 21st century we can still be moved by rituals and services that rouse an enigmatic primal call. It seems easy to imagine that our Stone Age ancestors would yield to that inscrutable appeal.

A Persuasive Leader

Sometime, possibly around 40,000 years ago, forms of rites or rituals were beginning to be overseen by leaders. Later, we would put our own labels on these spiritual trailblazers: shamans, medicine men/women, priests or priestesses. This type of leadership might have started from the practices of the burial of kin, linking the leader of the ritual to the dead loved one. The leader inevitably becomes associated with the dead and the mysterious force, which might only be accessible through him/her.

The leader would speak some form of proto-language to interconnect all the symbolic aspects of death, kin, and gods. Until there was an understandable, particularized language, the leaders who ruled the hunt would have wielded the most important form of power.

Even 40,000 years ago a type of certitude, along with a talent for convincing others, would have been needed for a religious/ritual leader. The complexity of human behavior within the tribal units at this point would have advanced enough that the spiritual leaders would not have needed to know how to track and kill big game animals. Their talent lay in a different direction.

Awe-inspiring ceremonies, sacred rituals, and animal sacrifices would have been overseen by the religious leader and carried out by the tribal members in return for the promise by the spiritual leader of taking over the mystery of controlling the uncontrollable.

The Next Chapter

The next chapter will discuss the Stone Age culture of 40,000 to 10,000 years ago. This is the beginning of Stone Age people's communication with the world. Paleolithic art is the material projection of groups of humans who lived in a very distant time. We'll explore the culmination of the Old Stone Age people's domination over the animals they lived among. The universality of their artistic expression makes it clear that for the first time, humans considered themselves separate from the other animals

Chapter 15: No Canvas Needed/Art in Caves and Art Artifacts

A hundred thousand years ago an artisan sat in a cave in what is now part of South Africa and, using a pestle, crushed material into a small abalone shell. The material was a blend of yellow ochre, marrow from animal bones, charcoal from a fire pit, and water. This now dried and dusty primitive blend, still extant today in its mortar and with its pestle, is the first evidence we have that humans purposefully mixed paint. Though the cave walls in South Africa where the dried paint matter is located are covered with thousands of years of encrusted limestone and do not reveal their secrets, the mixture was undoubtedly used for decorating either the enclosing space of the cave or the humans themselves; perhaps both. The cave is Blombos and sits on cliffs 180 miles east of Cape Town. The primitive paint tool kit reflects the deep roots of the impulse for humans to create some form of art.

Since prehistory, art has played an important role in human society, speaking to our ancestors in a vocabulary that resonated with them. The yet to be discovered traceable starting point for art probably dissects the point with which humans emerged. From the beginning humans found methods for communicating ideas in their art which were incommunicable before its development. An artwork speaks a language of its own: it is the visual thought of an artist, and that art-thought is communicating something to everyone who sees it.

Just how far back the invention of art extends has been the subject of much research and debate among paleoarcheologists and paleoanthropologists. Recently, the ancient abstract markings on cave walls in Spain have created a new dispute: exactly which species first created art? Thankfully, researchers are not yet arguing that squirrels or birds created the first art, but rather about whether *Homo sapiens* or Neanderthals were the first to set paint to the rocky walls in dark, deep caves.

New Discoveries of Old Sites

There are three newly discovered caves in Spain, each separated by several days walk, that have been found to contain what we would call art. One of the caves contains purposeful strokes of symbolic patterns of red ochre on its walls. These patterns are undeniably the work of a focused artist with a skill at creating what we would now call "abstract art". Most of the art on the cave walls dates back more than 65,000 years ago.

The newly discovered art works that have piqued the imagination of many researchers are:

La Pasiega, a cave near Bilbao, containing a remarkable ladder-like, abstract painting that has been dated to more than 64,800 years old;

A cave named *Maltravieso*, in western Spain, that has a hand-shaped outline that is thought to have been created by spraying paint from the mouth over a hand pressed onto the cave wall. This hand shape is at least 66,700 years old;

Ardales cave near Malaga in Spain, where cave stalagmites and stalactites have been painted with red ochre and have been dated to 65,500 years ago.

The Oldest Abstract Art

Today we could call the creations from the first two caves abstract art. Looking at photos of the newly discovered cave art, it isn't difficult to think that there will be some enthusiasts who will label these as long-lost masterpieces. It is possible that if the cave art had been painted on a canvas during the time of the great 20th century art movements of non-figurative and nonobjective images, the paintings would have competed with the best of our contemporary works.

At the height of its popularity in the 20th century, abstract art was sought after in its most uncontaminated form. How much more "pure" can art be than 65,000-year-old unassuming cave paintings produced by arguably the originators of the genre? If only the cave artists had lived in a recent time zone, by now their paintings would be displayed on the walls of modern art museums alongside the paintings of the great abstractionists like Paul Klee and Wassily Kandinsky.

So, is this newly discovered cave art another feather in the cap to show off the genius of modern humans and their ever-expanding human brains? Not exactly.

How the First Art May Have Been Snatched Out of The Hands of Modern Humans

The history of art has always been the domain of *Homo sapiens*. In fact, it has been evidence of our ability to think symbolically and held up as proof of the cognitive superiority of modern humans – examples of the exceptional skills that define our species. By comparison, Neanderthals, who we now know had much to offer, have suffered bad press.

Neanderthals were already making their home in Europe well over 200,000 years ago. Skeletons, tools and other remnants are found in European Neanderthal sites that predate the arrival of modern humans by thousands of years.

More than 65,000 years ago, some artist reached out and made designs with red ochre on the walls of a cave located in what is now Spain. But here is the rub: modern humans left Africa and arrived in Europe less than 50,000 years ago. Scientific measurements from all three newly discovered caves reveal that the art work and the paintings on the cave walls predate the arrival of modern humans by thousands of years. At this point we can give a round of applause for Neanderthal artists! They have been badly maligned, overlooked, and perhaps copied by no other people than ourselves.

The First Artists?

Cave paintings are not the oldest examples of art. There is a piece of 100,000-year-old red ochre carved with zigzag lines found at Blombos cave. The artist who carved that piece of hard ochre was a *Homo sapien* and lived in what is now the country of South Africa. Even further back in time, 500,000 years ago in Indonesia, a *Homo erectus* artist carved crisscross lines on a shell.

Cognitively, there is a portion of the human brain which may well be devoted to the element of creativity. Creativity is expressed in many ways, sometimes in spectacular paintings of wild animals on ascending cave walls.

France and Spain: Where the Master Artists Hung Out

The hunter-gatherers in France and Spain often made their shelters on overhanging cliff ledges, close to cave entrances. Their structures, starting 40,000 years ago, were made with wood and animal skins and were lean-to huts placed against the sheltering cliff rock walls. These were our Cro-Magnon ancestors, modern humans just like ourselves, with not only the same physical appearance but also the same mental capacity. These are our brothers and sisters. And some of them seemed determined to pursue a career in art.

Artists have always been considered a little different from the typical person. This might well have been the case 40,000 years ago when the early artists ventured out of their lean-to huts and into the caves in the hillsides in France and Spain. They sometimes crawled through narrow, muddy passages for hundreds or thousands of yards.

The cave entrances were often narrow, and the cave passages varied in degree of difficulty. Often the rock floors were slick and treacherous. Sometimes the passages themselves were narrow, and the primordial clay underfoot oozed with muck. Slopes of the cave passages made it very awkward to access the inner sanctum. Frequently the only method was by sitting and scooting while steeply descending to the bottom. It was when the cave artists found the looming, massive gallery rooms that they stopped and created their magnificent paintings, engravings, and bas-relief sculptures.

The caves which hold most of Stone Age art were never the domiciles of the hunter-gatherers. Caves are deep and dark and cold. To start a fire to warm an entire tribe would quickly be met with a suffocating accumulation of smoke. Instead, the caves were special places, dedicated to the expression and possibly the celebration of symbolic imagery. They were what we might call cultural sanctuaries.

Altamira: Spectacular Paintings

The artists who painted in caves like Altamira (Grotte d'Altamira) in Spain, had visions of communicating their art that were quite different from the abstractions of the Neanderthal artists. At the time the paintings were created, as far back as 35,000 years ago, the entrance to Altamira was probably not easy to enter. Presumably the cave interior was what interested the artists who chose Altamira. The inner chamber with its deep gallery (the "Gran sala") was possibly selected so that many hunter-gatherers crowded together could fit into the basilica-like space with its soaring ceiling.

For the uninitiated tribe members, the anticipation of viewing the paintings must have been exciting. Once inside, carrying animal-fat-lit, small stone lamps, they must have marveled at the great painted ceiling, which is the vision that dominates Altamira. The ceiling is covered with vibrant, dramatic paintings of almost life-sized bison that appear to be plummeting across the sky. The bison make a powerful visual impact. For the contemporary viewer the immediate impression is that this is accomplished art.

Though the paintings at Altamira are dominated by the eighteen almost life-sized bison, there are several other animals represented in the cave; two horses, a wild boar, and a female red deer. The deer is conspicuously the largest single animal painting in the cave. All the images are painted with the skill that is required of any artists of representational art and are so convincing that after a few minutes it is possible to think that the humans viewing the art thousands of years ago thought they were the real thing. Indeed, Altamira has been nicknamed the Sistine Chapel of Cave Art.

Most of the herd of bison are standing, but a few of the natural indentations in the cave ceiling made it more convenient for the artists to paint the bison curled-up in the cave wall contours. These are the images that appear three-dimensional.

Carrying the small, almost flat lamps, the tribe of hunter-gatherers, (to be very casual with the term we could possibly say "the museum visitors") were sometimes able to see the paintings in a dream-like setting. The wavering flames of burning tallow played tricks with the light in the cave. When groups viewed the paintings together, the many tiny flames reflected on the projected rock shapes made the paintings come to life. As the hunter-gatherers moved through the cavern, many of the bison would momentarily look as if they were swaying, expanding, or briefly disappearing. The illusory motion was the effect of the play of shadows that distorted and animated the paintings. It is certain that the "moving" creatures would have been as impressive to Stone Age people as the newly invented "movies" originally were in recent times.

The great representational art of the ceiling paintings of Altamira did not happen all at once, but instead happened in phases. The earliest art was decorative and engraved line drawings on the rock walls. Then, later in time, figures were inserted in a red flat-wash. Eventually some black figures were added to the mix. Later still, and lastly, around 17,000 years ago, the famous multi-colored (polychrome) bison and other animals were added.

Artistic Techniques in the Time of Stone

The Stone Age artists painted with different solutions of water, charcoal, manganese and ochre, diluting and mixing the black and reddish-orange pigments to create different color intensities. The cave pigments were applied in a variety of ways. The most "contemporary" manner was with a brush made of hair from the tail of a bison or horse. Visible brushstrokes are still discernable on certain wall paintings. Pigments were also daubed on with sponge-like materials, such as dried moss or fur. Paint-dipped fingers were frequently used on the animal figures to create shadows and darkening.

A technique where the liquid pigment is blown or sprayed was a very popular procedure for the cave artists. Finely ground pigments of color mixed with water were blown through hollow bird bones directly onto the walls. This blowing technique requires that the paint be held in the mouth and then applied to the wall. The other method, just as popular, was to blow the color directly from the mouth onto the wall without the aid of a blowpipe.

Outlines of eyes, horns, legs and bellies were applied with sticks or brushes dipped in manganese to highlight and define the details of the animal.

Amount of Preparation and Planning

Since the artists had only one source of light which came from the small stone lamps burning animal fat, they had to carry enough fuel into the pitch-dark cave to last their entire trip. Time had to be calculated to include the interval spent crawling, scooting and walking in and out of the cave, the painting of the walls, and performing other tasks. A sudden unexpected draft could blow out their lamps. Dropping the lamp in a moment of disorientation would also extinguish the lamp. Incidents must have happened.

The more difficult thing would be for a single artist, looking for new walls to paint, striking out on their own, venturing alone into the deep fingers of the cave's many tunnels. Stuck in the subterranean depths with a single lamp, any mistake and the tiny flame could be extinguished. Swallowed up by darkness, the noises, echoes, and water splashes might seem to magnify their isolation and the ominousness of their predicament. Yet it doesn't appear that any of the artists got lost; witness the fact that no skeletons have been found in the caves, with or without an artist's beret.

If you are thinking the art work must have required a great deal of preparation, you are right. The art not only necessitated the skill of a talented artist or artists, but also numerous helpers. In many of the caves, the paintings are very high on the walls, or, as in the case of Altamira, on the ceiling itself. Firm scaffolding would have had to be set-up wherever the artist or artists were painting elevated pictures. These would be spaces on the cave walls that would otherwise be inaccessible.

Before the project in the cave could begin, preparation would be carefully planned. With more than a little effort, wood for the scaffolding must have been carried in. Tree branches would have been selected, cut to size and then dragged in through the cave entrance. The paint itself, once applied, was not erasable. The artist would not be able to wash off or get rid of any of the images. Since all corrections would show, the paintings had to be as perfect as possible.

The amount of forethought and arrangements before a figure could be painted on a cave wall makes it highly unlikely that it was ever only one or two artists who were involved in the creation of the cave art. Instead, the painting of the art must have been a social affair that included many in the tribe and looked over by one or several master painters. There might have been some sort of a school of art at this time which selected the more gifted in the tribe for training. There had to be a few true genius artists who were supported by members of the group. Their role would probably have been limited to create and work on the paintings. The role the paintings themselves played in the culture of the Late Stone Age is the stuff books on cave art are made of.

The Theory of the Hand

One of the ways that today we can be in direct contact with the cave painters is to look at their handprints. Handprints occur with regularity on most of the cave walls. There are two ways to create handprints, a negative handprint or a positive handprint. The cave artists used both methods.

A negative handprint is created by placing a hand flat on the cave wall and then pigment is sprayed on by either a blowpipe or directly by mouth. This will leave an outline of a print with a cloud of color surrounding it. The alternate method is a positive handprint which is made by pressing a paint-covered palm against the cave wall.

There are many theories as to why the handprints consistently occur in scattered spaces between the animal images on the walls of the various caves. One theory is that the artists were using the hand as a proportional devise to measure the relative size of the animals they were drawing.

Another theory is that the handprint represents human intelligence and control. It is the human hand that creates the tools, blades, spears and weapons. It is the hand that throws the weapons. The hand, in this theory, is the symbol of the dominance of humans over the animals. It is the hand that represents which animal will live and which will die.

Still another theory of the handprint is the spiritual aspect of its image. Historians point to the touch of God which is a familiar theme in Christianity. God touching the hand of man is the principle focus for a famous painting in the Sistine Chapel. The positions of hands and fingers in artworks and sculptures are significant not only in Christianity but also in Buddhism and Hinduism.

Lascaux: Breathtaking Cave Art

Lascaux cave is in the rolling hills which surround a green and lush valley. The serene river of Dordogne that flows along the valley floor creates such a picturesque sight that the area is often called the "country of enchantment". The most significant ancient history in the Dordogne is its painted caves. The famous Cave of Lascaux (near Montignac) contains one of the world's most extraordinary repositories of prehistoric wall paintings. Though the area of the Dordogne, with its fertile river valley, is riddled with prehistoric caves and their paintings (*grottes ornees*), some dating back 30,000 years, Lascaux is its most remarkable.

Lascaux cave is best known for its 600 magnificent paintings of aurochs, horses, deer and geometric signs. It also contains almost 1,500 engravings of horses (for the most part). The paintings seem to live and be integral to the cave walls. Like Altamira, the artwork is representational and recognizable. The animals run, jump and tumble, as if the artists were fully aware of the spectacle they were creating. As was the case with Altamira, Lascaux was painted in stages, probably taking hundreds of years to complete. The latest and most magnificent of the paintings date to 18,000 years ago.

Lascaux's artists were extremely adept at capturing the vitality of the animals they depicted. They did this by using broad, rhythmic outlines which frame areas of soft coloring. Frequently, animals are depicted in a slightly twisted perspective, with their bodies shown in profile but with their head, horns and antlers painted from the front. The result is to imbue the figures with a personal identity, one that the viewer can relate to.

The Paintings in Lascaux

Lascaux offers an amazingly diverse repertory of images. For the most part, it features the traditional animal paintings of Ice Age art, composed in an original and enchanting way and sometimes incorporating graphic techniques as well. Among the large herbivores represented in the paintings are the aurochs, extinct relatives of modern cattle, which appear nearly everywhere throughout the cave. Bison are less common, while horses are omnipresent. Often all these animals are portrayed in groups or in a line, creating the idea of a herd.

Though reindeer were stalked, killed and eaten by the Ice Age hunters, there are no images of reindeer in the cave art at Lascaux. But images of red deer abound. Contrary to most of the realistic images of animals in Lascaux, the exaggerated and deformed antlers of the red deer make them look, in places, like invented beasts. Yet the ibexes, a close relative of the red deer, are portrayed realistically with clear-cut contour lines to illustrate their large curved horns.

The images the artists painted reflected the world that was important to them. The artists were not attempting to re-create and record the actual domain they lived in. Nor are the images painted on some whim. Instead, they chose to paint animals that had a special place in their society. As with the case of the reindeer, the animals were not necessarily the animals they pursued in the hunt, but probably rather the ones that were valued in an aesthetic/mythological or spiritual way.

The Great Hall of Bulls

As opposed to other larger painted caves in the area, where the images are painted deep in the cave galleries, Lascaux is unique in that the spectacular images appear immediately after passing through the entrance. Once moving past an enigmatic and imaginative painting of what looks like a mythological unicorn, the Hall of Bulls appears. It is thought that 18,000 years ago, the wide entrance to the cave was a smooth slope leading to the Hall of Bulls, and that daylight illuminated the figures.

The Great Hall of Bulls of Lascaux contains several immense auroch figures. One of the painted auroch bulls is over 17 feet in length. This is the largest painted figure of known Ice Age cave art.

As the hunter-gatherers filed into the Great Hall of Bulls, they were able to see an uninterrupted line of painted animal images, all created to seem in motion. Several horses in dynamic poses follow the Unicorn. A large, two-colored horse comes next, followed by the first of the many aurochs. Closely following the aurochs are four red deer. Finally, an enormous image of a red bison is situated to the right. The animals are surrounded by many geometric signs and are placed high on the wall, six and half feet above the current floor.

Geometric Signs

There are several types of mysterious, geometric symbols at Lascaux. These marks or symbols are emblematical. For the most part the inexplicable symbols are limited to large rectangles, straight lines with side branches, star shapes, or rows of punctuation dots.

The many geometric signs in the tableau of the Great Hall of Bulls are arranged on and around the animals and for the most part are multiple versions of complex star-shapes. This graphic star symbol is also engraved as decoration on spear points made of reindeer antlers that were discovered on the cave floor.

As the hunter-gatherers walked through the Great Hall of Bulls they would have seen that the painted walls and further into the cave, the ceiling, are continued on into the next room. The high and hard-to-reach domed vaulted ceiling is covered with animals and signs. The signs are dotted around the animals as if they are scattered randomly, though these motifs might not be random; their purpose in the paintings is not known. Some of the graphic elements are simple—single dots, a series of dots, black rectangles, crosses, hooks; other signs are more complex and are grouped together or linked to the nearest animal.

These symbols in the cave art are code, though created without the printed word. There is no Rosetta Stone to translate these prehistoric messages and they remain indecipherable. Yet it seems clear that the codes are in some way specific markers or signatures for the group and are intimately linked to the animals, and so also, to the culture that created them.

The Falling Horse

Beyond the Hall of Bulls in another of the many galleries, is a remarkable painted figure called the "falling horse". It is painted around a rock in such a way that the artist could never have seen the whole figure at once, yet when the figure is flattened out with photographs, it proves to be in perfect proportion.

Human Figures and Other Carnivores

Another cave gallery within Lascaux features a painted scene of a bird-headed man with a wounded bison. This is a narrative scene which is more easily interpreted. The man has wounded the bison, whose intestines are spilling out, and the bison has knocked over the man who is lying on the ground next to the wounded bison. Immediately to the left of these two figures which are locked forever in mortal combat, stands the only image of a wooly rhinoceros within the walls of Lascaux. Beside the man who is lying on the ground, probably also wounded, are two sticks, one with an image of a bird at the end of the stick and another stick with a barb at each end. The man's bird head, his hands that resemble bird's feet, and his stick with a bird at one end, have been interpreted in shamanistic terms. Shaman, who bring magic for hunting, would be sought-after leaders.

The image of the man/shaman, which looks far from representational, is drawn in a child-like, stick-figure way. He is a figure that has a long torso where the chest is missing as are the forelegs. Symbols of two rows of black dots, each row consisting of three dots, are painted immediately behind the wounded bison and the fallen man. These dots, along with the other numerous symbols in the cave, are waiting for an interpreter. If you have any ideas, contact the Minister of Culture in Paris.

In another narrow part of the cave is the "Cabinet of the Felines" that is filled with engravings, including a remarkable horse seen from the front. The most dominant figures are a group of realistic looking lions that is clustered in a narrow passage at the end of the cave.

Sex and Birth

There is an unsullied nature to the cave paintings, an incorruptibility that is somewhat surprising considering that they were painted over 18,000 years ago when people, presumably, had intimate visual carnal knowledge of each other. At least, considering the flimsiness of the animal skins of their huts, privacy was at a premium. Yet, within the walls of Lascaux there are no pictures of animals or humans mating. And with one exception, no animals are giving birth. Images of vulvas, penises or pregnant women are limited to geometric shapes that are more of a suggestion than they are pornographic drawings.

Empty Space

There are many things that are left out of the paintings of Ice Age cave art. There are no landscapes. The animals exist free from rivers, cliffs, caverns, or rocks; free from nature. There is never an image of a tree or a flower or a bush. The animals might be standing, running with the herd, fighting, or in some cases, falling. Yet, there is nothing to hinder them. The skies, the sun, the stars, the moon, are all insignificant. Natural surroundings have disappeared.

Since there is a popular theory that the cave paintings were painted as some form of worship, the exclusion of the skies above is a puzzling omission. The movement of the seasons, the migrations of the animal herds, even their own tribe's migration, all would have been marked and anticipated by some form of astrological observation.

Other Caves: The Sorcerer

Les Trois-Freres is a cave located in southwestern France famous for some unique cave paintings and engravings. Most of the art work is about 15,000 years old and, in its style, it is associated with the other cave paintings in France. What makes the cave of *Les Trois-Freres* so different is that it has one of the most distinctive painted human figures ever found in the caves. The whole figure, which is a combination part-man, part-beast, is 30 inches tall. He is shown standing on his right leg with his left leg lifted and moving forward. This gives the impression he is moving fast, possibly dancing. The name given to this creature is "The Sorcerer", though he may also represent a shaman, or a depiction of how the hunters disguised themselves when stalking bison.

The upper half of The Sorcerer is bewildering. It is either a picture of a man with a bison mask placed on his head, or it is meant to represent a mythical bison-man. His eyes are two black circles and his head is turned directly to the viewer. He is looking straight at us. His nose is a single line between his eyes that ends in a small, down-curved mouth. A long, pointed beard covers the rest of his face and reaches to his chest. Two large curved horns stick out of his bison head and his ears are like those of a stag's. He has muscular, human legs. His belly is hairy with a tuft of fur at his navel. A first impression is that surely this bipedal creature is related to neither kith nor kin; a creature so strange that no human would claim him either then or now.

There have been many interpretations of the paintings in the caves of Spain and France.

Possible Meanings for The Cave Art

Theory of Art for its Own Sake

We who love art might at first naively think that the paintings and drawings on the caves were created simply as "art for art's sake". Though it can't be entirely ruled out, it seems unlikely that the hunter-gatherer tribes, who lived in harsh conditions and put in enormous effort to get their food, would put up with several dozen of their tribal members joyfully painting away in the depths of the local caves. The cave art at most sites has been very carefully designed to communicate a story or message. The idea that only beauty was on the minds of the cave artists would be to underestimate the powerful meanings that the art conveys.

The hunter-gatherer artists were careful to choose only specific animal subjects, not, for instance, any other part of nature like landscapes with mountains and trees. The cave artists often ignored painting certain very common prey animals in favor of other animals. Many times, the artists chose to paint lions, for instance, instead of reindeer. The argument that the hunter-gatherer artists only painted things because they were beautiful cannot solve these conundrums.

Theory of Magical Visualizations

Another theory offered as an interpretation of the Stone Age cave art claims that the artists were tribal shamans and were trying to reproduce the visions they saw while in a magical trance. The theory maintains that their drawings and paintings of animals were to put a pictorial incantation on the real animals and by magical suggestion put the animals under a spell so that the tribe could easily hunt them. In this theory, artists painted pictures of wounded bison in the hope that this type of primitive "visualization" might make the bison become wounded.

Theory of Spiritual and Nature Connections/ Cultural Overtones

Arguably the most convincing explanation for the cave art is that the paintings were created as part of some spiritual rituals. The theory is that realistic cave paintings found throughout Europe, some dating as far back as 45,000 years ago, represent the earliest evidence of human spirituality and demonstrate our ancestors' interdependent relationship with the creatures they hunted. The great master cave artists were also the apex predators; their intimate knowledge of the physical and behavioral facets of the animals they portrayed demonstrates how carefully they were acquainted with their subjects. There are no images of domesticated pets; no fuzzy, ever-so-cute lap dogs or fluffy cats painted on the limestone walls.

Yet, the animals on the cave walls and ceilings are rendered in a way that is at once realistic and sympathetic. There are images of both herbivorous and carnivorous animals on the cave walls in this, the end of the last extreme cold period of the Ice Age. This theory suggests that the art in the caves expresses the spiritual closeness that existed between the humans and these animals.

Sanctuaries and Ceremonies

At the heart of the hunting societies, cave art accompanied and perhaps supported a multitude of social practices. The caves were most probably used as religious sanctuaries for ceremonies; some of the ceremonies were undoubtedly initiations for hunting rites. Several foot imprint studies have shown that virtually all the footprints in some of the small ancillary rooms of the caves were left by adolescents; a typical category for initiates. The seclusion and isolation of the caves, with their painted walls and lighting created by the hand-held flickering flames of small stone lamps made the caves an ideal place to conduct ritual ceremonies.

The First Rock Music

Thirty-five thousand years ago ceremonial dance is thought to have been accompanied by music. Archaic people had various forms of percussion instruments, created by hitting rocks or wood against hollow objects. Clapping, singing, and other forms of instrumentation were also undoubtedly used.

Ancient flutes have been discovered in Paleolithic sites around the world, from East Asia to Europe. There have been Paleolithic musical instruments found in caves, on ledges and in living sites dating as far back as 45,000 years in Germany, France, Turkey, and India. Most of the flutes that have been found were made of hollowed bone, incised with stone tools to form air holes that changed the pitch. The recycling of the hunt might have included bones that would have been available after a meal. Bird bones were used as flutes, as well as the bones of deer, cave bear and ivory mammoth tusk.

The oldest of the instruments found so far is a flute that is 8.5 inches long and was made from the bone of a vulture. It has five finger holes and when researchers made a wooden replica, the instrument played a range of notes like modern flutes. This gives a special meaning to The Sorcerer cave painting, creating a mental image of a dancing shaman.

Yet, though the musical instruments of early humans show that they were steeped in song and tunes, the Paleolithic music itself can only be imagined.

Acoustic Properties

Some of the art painted deep in the cave systems was not meant to be viewed easily. There are many hard-to find passages in the caves and the effects on the uninitiated coming into hidden grottos painted with realistic or imaginary figures must have been deeply impressive. The cave artists were masters of their craft, certainly aware of the impact of their images. The ceremonies that must have been carried out in the caves were accompanied with songs, chants, music and possibly noises such as animal calls. The chambers of some of these more hidden cave rooms have been tested for their resonance and they appear to have had special acoustic properties.

During the Paleolithic era, the Ice Age animals nurtured the people and added to their culture. The animals they painted were beasts of the field, some were the herd animals, and some were apex predators like the humans themselves. The people of the late Ice Age were inspired by the presence of the animals they lived among. That inspiration stimulated their imaginations and spoke to their hearts and minds. The cave art is the result.

Portable Art

Personal adornment started over a hundred thousand years ago with perforated shells, canine teeth pendants, and carved bones. Neanderthals and perhaps *Homo erectus* were not without their interest in small objects that were decorative and easily carried. Dyed and decorated seashells made by Neanderthals have been dated as 115,000 years old. The genus *Homo* seems to have had a long artistic tradition.

Cro-Magnon Portable Art/40,000 Years Ago

Besides the earliest appearance of Cro-Magnon cave art, the dawn of elaborate artistic expression also included carved statuettes, decorated pendants, beads, and designs on diverse loose material. Modern people, who entered Europe 45,000 years ago, were working on their artistic skills. For instance, forty thousand years ago a figure of a woman (*The Venus of Hohle Fels*), carved out of the tusk of a wooly mammoth and only 2.5 inches tall, has been found to have been worn as an amulet.

It is thought that small objects that could be carried around and worn were used as a sort of traveling art that probably served various purposes. The new technology of artistic decoration is thought to have come out of the flint technology of the hunters who decorated their blades. For the first time, starting approximately 40,000 years ago, bone, ivory, and especially reindeer antlers were chosen as raw materials for hunting weapons. The artistic articles of the first modern humans to reside in Europe were fashioned with these same materials. The type of work that was created by the modern humans is called "the bone industry".

Decorated ornamentation from this, the people at the end of the Ice Age, consisted of a multitude of beads, bracelets, necklaces, and pendants, made with antler, shells, animal teeth, marine fossils and various types of stone. These objects were all portable and unlike the contemporaneous cave art, could be carried by the individual wherever they went.

Portable art also included statuettes of all kinds, many of them only a few inches high. These were fully-sculpted, three dimensional figures that were carved into ivory, antler, stone or bone. There have been several dwelling sites located near the caves where there was a virtual industry of portable art going on. Adornment with beads made from the teeth of fox, reindeer, stag, and bear have been found by the hundreds at those sites. The sites might have been the Tiffany's of the Late European Stone Age.

Signs of Early Trade

The numerous items of exotic artifacts are a testament to how valued the adornments were. The shells were carried from the Mediterranean or even the shores of the Atlantic Ocean; the ivory was brought from distant mammoth-populations in Eastern Europe. It can't be ruled out that trade was one of the reasons that material from distant places was found in these dwelling sites. Culture was beginning to play a prominent role, whether the goods were traded for barter or used as matrimonial settlements, or ritual initiations.

Gender and Art/ the Lion Man or Woman

Anthropomorphic figurines from this period are the examples that are most familiar to contemporary readers. The figurines were carved as both male and female statuettes. The most famous of the male figures is the ivory carved "lion man". The lion man is the oldest-known example of figurative art. You will find a photograph of him in most art history text books. This figure was found in a cave site in Germany. It has been dated as being 40,000 to 35,000 years old. The body is human, but the head is that of a lion. It was carved out of wooly mammoth ivory using a sharp flint stone knife. The genitals have gone missing which is a sad story in more ways than one. Since the lion-head on the statue has no mane, women have claimed that it is not a lion man at all but really a lioness. However, European cave lions didn't have manes, so the sex of the statue has not been determined. The controversy rages on with one museum official in Germany stating that it has become a shibboleth of the feminist movement.

Venus Figures

The most famous portable art of this period can be unequivocally claimed by women. These are the abundant Venus figurines depicting stylized, three-dimensional figures made of ivory or stone. A variety of types of these female figurines were available in the late Stone Age for tenderly holding. Their bodies range from slender to fat, from children to grown-up, and even a few very old women. Most of the Venus figures are portrayed with large breasts, and rounded bellies and hips. Some of the Venus figures are clearly pregnant. The bodies are rendered in a formalized manner: voluptuous yet lozenge in shape, with relatively small arms and legs, legs ending in a point with no feet, and no personified facial expression.

An amazing feature of these statues, found all over Europe, is that there seemed to be an understanding by the many artists who carved them that the statuettes were to be rendered only as symbolic forms and never as realistic versions. The detailed abstracted bodies of the statues are calculated to the point of being an artistic convention. This indicates an understanding among vast numbers of Stone Age people that the Venus figurines were developed and created for a shared, codified significance.

The most famous Venus statuette is the Venus of Willendorf. The little figurine was found on the banks of the Danube River in Austria. It is one of the earliest images of a human body. It stands just over 4.5 inches high and was carved over 25,000 years ago. The exaggerated image of this female figure shows the breasts, belly and hips of a very plump woman. Her arms are disproportionately short for the rest of her body. Her large thighs converge into extremely shortened legs. Her round head is decorated with rows of prominent bumps, though these bumps are symmetrically designed and probably represent a headdress. The bumps also cover her face, making her facial features, like all the typical Venus figurines, unrecognizable.

When the Venus of Willendorf was made 25,000 years ago, the environment was very cold and harsh. The Ice Age was slowly coming to an end, but it would not vanish until thousands of years after the artist who created this sculpture had gone. The meaning of the Venus statuettes is contested, but fatness was probably linked to fertility. The most popular interpretation for the Venus figures is that they are symbols of fertility or mother figures. The female figurines could have been used as charms for successful hunting or as a talisman for fertility for women or as symbols of fertility for the land.

Although there was no Stone Age written record of what the figurines were used for, the Venus figures were named for the Roman goddess of love and sex. In Latin, *Venus* means love and sexual desire. Very few people in the world today or any time, would look at a statuette like the Venus of Willendorf and interpret it to mean "delicious wooly mammoth soup". Even scholars, not known for their passion and amorous desires, have concluded that these figures represent fertility. The degree of significance is the only question: was the use of Venus figurines created for charms or magical potions of some kind, or were they representations of an earth goddess, and symbols for an entire cult of people who believed that women were the rulers of men and animals?

Chapter 16: Are We There Yet?

We have come a long way since the Stone Age. But how far have we come? For instance, climate change is not new. Our Stone Age ancestors trudged through snow and ice, or at other times, through drought-ridden landscapes filled with sand. Some change will always be inevitable, whether it is in our climate or within ourselves. Our coping strategies seal our destinies. Our species is now preoccupied with many things, most of them not having to do with observing or considering the natural surroundings in which we live.

The early members of our species probably thought little of their future; they were tending to the present, often harsh conditions in which they found themselves. From small, tree-dwelling ape-like creatures to the first up-right, bipedal, vulnerable humans walking the muddy earth at Olduvai Gorge, our situation did not augur well for our future. Genetic changes aside, there must have been a large dollop of luck involved in our success: somehow a not-too-unique and not-too-superior organism turned into the exceptional genus *Homo*.

Though we do not usually think of it, there was a good possibility that our genus would not have materialized at all. Had there not been just a certain combination of climate, land plants, animals and lack of catastrophic threat from other species, it is likely that some creatures other than us would now be ruling the earth and seas. To be born at all is a very great privilege. The large part that chance played in our presence on earth should give us all an immense feeling of humility.

<u>Forgoing Many Things</u>

Whether the species of *Homo sapiens* was one of a series of evolutionary accidents or an integral part in a universal plan, the process of natural selection required some deep sacrifices. Childbirth is accepted as a given fact; what alternative do females have? Bipedal women with their functional walking legs and consequent narrow pelvic canal meant that giving birth was a major sacrifice for the human species. An infant, in utero, with a large head that holds a hefty brain means that during child birth, enduring pain was inevitable. The helpless offspring comes out at great cost to most women, but the child has a workable brain that, added to the collective human intelligence and knowledge, hopefully will be clever enough to help ensure the survival of our species.

The Luck of the Hairless

Setting aside the limitations of the human body, there were times when it looked as if there would be additional reasons why a species other than *Homo sapiens* would be the inheritors of the globe. The original range of our territory was not nearly that of *Homo erectus*, a species of hominins who took off from Africa thousands of years before us. *Homo erectus* had bodies closely resembling modern humans', could make tools, used fire, and possibly were the first to use boats.

We *Homo sapiens* were hairless, lacked any means of independent food production, and were at the mercy of an unstable planetary climate. It didn't seem likely that our species would emerge from Africa, dominate and eradicate other species, become the apex predators, progressively establish settlements on every continent on earth, and eventually learn to fly off the planet checking out the possibilities of colonizing other worlds.

We have learned much. But have we learned enough? Looking at our ancestors' lives and activities forces us to reconsider long-held assumptions and ideas about our present behavior and our future.

Hard-wired Instincts Inherited from Long Ago

Small preoccupations inherited from the past play a larger role than we might imagine. For instance, there was a fixated search for more fat during a time when our hairless and not-well-clothed ancestors needed internal warmth. Today our craving for fat and sugar (in the case of archaic people...honey) is inexorably linked to our primitive past in a way that means we need to take a careful look at these ancient internal cravings. If we face the basic desires, we have a better chance of keeping them under control.

Many hard-wired instincts from the past are still indispensable. Some of these instincts we need almost as much as our archaic ancestors. The ability to size up a situation, think quickly and make rapid decisions can be as necessary today as it was in hunter-gatherer times.

An essential requisite for our species is pair-bonding, allowing the lactating mother with her dependent infant to be helped and protected by her mate. Without the supportive tie of the unencumbered male who can hunt and bring food to his family unit, the species would have long ago suffered calamity. Apparently, bonds such as marriage were originally based on the primeval notion of the survival of the species.

Gossip appears to be a hard-wired behavior. In the Stone Age gossip was used for regulating and controlling the tribe. We who live in contemporary times use our electronic media to effectively behave in the same manner. Though it must be admitted that today there are aspects of gossip and shaming posted on social media that would make even a Stone Age person blush.

We may be far more uncompromising than we are willing to admit. A question that is difficult for we subjective humans to face is whether natural selection has fixed in our genes a certain model of behavior that we seem unable to banish. Just how aggressive and competitive are we? Hard-wired tribalism is not an historical, distant chronicle from the past, but a factual truth evidenced in all the current wars around the world. Most people still suffer from generalized, deeply entrenched biases.

Could we ever agree to eliminate our intertribal wars? The aggression of our species toward Neanderthals may or may not have been the reason for their extinction. But aggression, which is not limited to, but is more prevalent among human males, certainly can lead to the death of many or even the annihilation of an entire species.

Fearfulness from Long Ago

Using the knowledge of nascent fears would be helpful when facing some obvious irrational panic attacks. We no longer need to be afraid of falling out of trees when we are asleep, the mattress is never that high off the floor.

Though many people in developed countries consider the possibility of wild animals attacking or killing them, statistically, the odds are so remote that this is an irrational fear. Yet, it was only 1975 when there was a national panic over sharks caused by the movie "Jaws". How many times in the life of someone who is reading this book have they been attacked or even threatened by a shark, venomous spider, or poisonous snake? These fears are holdovers from a time long-ago.

Shamans of old, dressed in frightening clothing, (think "The Sorcerer" in the cave painting), danced around the sacred space of the cave probably to motivate the members of the group into believing in shadowy apparitions. Today, we have our own shamans, still ready to tackle the invisible forces and purge the devils within. Fright-night movies that took their clue from "The Exorcist" are still there, ready to instill terror in the viewers. If we understand that these fears come from the primitive and long ago, they are more easily handled.

If we could face the fact that some of our complex, deeply emotional conduct is impressed in our brains from a long ago past, more headway could be made for a thoughtful, peaceful and cooperative future.

The Shared Aspects of Our Behavior

Some of our uniquely human behavior turned out to be shared with other species. Going back in time even farther than *Australopithecus* to our distant chimpanzee cousins, we see that they too use tools, live in social communities and express emotions.

The connection between what was thought to be distinctive about *Homo sapiens,* their burial rites and symbolism, has been expanded to include Neanderthals and possibly *Homo naledi*.
What was once considered under the aegis of human traits, their ability to create works of art, is now seen as an attribute that *Homo sapiens* share with Neanderthals who painted symbols on cave walls thousands of years before us.

If the social identity theory is correct, the roots of our morality go far back to our early primate legacy. This theory suggests that our concept of who we are and what we believe is shaped by the groups we identify with. This often leads to behavior that imitates and follows the mind of the crowd over individual choice. We share this characteristic with the great apes. In the case of humans, much good and much bad behavior can be explained by our primal urge to conform to the group.

Coping Strategies

There were some coping mechanisms that our species learned long ago:

Cooperation for the good of the group was accomplished in many ways in the Stone Age. On a regional level, the highly mobile hunter-gatherer tribes could clear out and leave if a rival group threatened them. It is a strategy for alleviating tensions that became much more difficult to achieve when humans discovered real estate and property rights. Yet the tactic of moving to reduce chances of violence stretches back tens of thousands of years. On a personal level this technique also works: if two individuals in a hunter-gatherer band are about to become aggressive, one frequently shifts to a different location.

In the contemporary world, cooperation is still an upmost necessity. Some of the palliative relief to reduce aggression between countries bound by borders goes back to archaic times. A technique for softening tensions between dissimilar groups has been to use trade. Interest in sharing and trading both utilitarian items (skins, furs and stone and bone tools) and decorative accessories (painted shells and Venus figurines) has the beneficial effects of promoting interconnections. Where trade goods do not cooperatively pass the boundaries, enemies might.

Females as the mate selectors have been with the genus *Homo* from very early on. This selection process is not limited to our species but instead runs through many animal groups. It is linked with Darwin's theory of natural selection. In today's world it might be good to remind females that finicky selection of a mate has its advantages and is linked with species survival. Of course, try telling that to a 14-year-old girl on her way out the door to meet up with her friends.

Hard-wired coping mechanisms that date back farther than our species are the closeness and compassion that bonded the early peoples together. Researchers have found evidence of early practices of morality among Neanderthals and *Homo sapiens*. Besides close connections to family and tribes, there are indications of care for the ill and physically challenged members of the group.

All these early coping mechanisms were learned over thousands of years of experience. Happily, there are many modern-day proponents practicing the same behaviors today.

The Primary Assumption for This Book

In this book I have gone on the assumption that the archaic humans that came before us were "almost like us". We gain perspective by peering over our shoulders to look back at them. This was the supposition I discussed in the introductory chapter. It seems to me that looking more closely at our past biological and cultural footsteps broadens our view. The story of human evolution illuminates how the complicated cathedral of the modern mind came about. How much of our human progress was of our own making and how much was due to nature is for you to decide.

In their time, our archaic ancestors pushed through and overcame many challenges. The strength of our ancient forbearers is something to be proud of. We have inherited their genome. We have their abilities and aptitude, but because of eons of collective learning and sharing, we also have a knowledge they did not and could not possess. The future, if we are reflective and forge on, looks positive.

Sitting on the ground and knapping away at amorphous stones two million years ago, *Homo habilis* took the initial actions that launched us toward our eventual rise to preeminence as a species. In both the literal and figurative sense, they were beginning to create a future for us with a sharper edge. It is up to us to continue the quest; hopefully we will come up with germane and thoughtful goals as we consider what we want our own legacy to look like.

"Only if we understand, will we care. Only if we care, will we help. Only if we help shall all be saved." Quote from Jane Goodall

Acknowledgments

My thanks to Ranny Eckstrom for her devotion and patient editing and for the constant stream of up-to-date relevant information that she sent my way. Thank you to Neita Gardner and David Lewis for their excellent help. As always, sincere thanks to my husband Jeff Hendy, a computer whiz and sometime researcher, who was essential for the writing of this book.

Chapter Notes

You & I and the Stone Age Guy

6   We all have two parents, and they had two parents, and all of them had two parents, and so on. There isn't a way to see this pattern doubling all the way to the Stone Age. Our heritage falls in on itself, which means all of us are enmeshed in a web of common ancestry. The 7 billion people who are alive today are descended from a handful of people from the past.

The Past World/Lives in the Stone Age

13   The Pleistocene (also known as the Ice Age) is the geological epoch lasting 2.5 million years ago to 11,000 years ago. This is the time of the beginnings of the genus *Homo* and spans the most recent period of repeated glaciations. The end of the last glacial period is also the end of the Paleolithic age (an archaeological term).

Climate Extinctions/Survival: Our Origin Ancestors

29   A "stone knapper" is one who works stones into tools.

30   Early human females probably gave birth every 3 to 4 years. Breastfeeding meant that the females did not ovulate and therefore did not become pregnant until the off-spring was weaned.

Knuckles on the Ground: But Not All the Time

33   Until the 1960s, toolmaking was widely considered something that only humans do. Then Jane Goodall observed chimpanzees stripping leaves off sticks to create a tool that fit in a termite mound. Defining human behavior becomes more and more difficult when one realizes that chimpanzees create tools.

34   Like great apes, *Australopithecus* had large canine teeth to show how big and tough they were. This is very different social behavior from humans. The large canine teeth signal threat when shown (think werewolf). The males probably competed for dominance by showing their teeth.

36   Laetoli Footprint Trail: A trail near Olduvai Gorge in Tanzania contains footprints that were probably made by *Australopithecus afarensis*, an early proto-human whose fossils were found in the same sediment layer as *Homo habilis*. The trail is almost 88 feet long and includes impressions of footprints of several early humans.

The Most Durable Species Ever: Homo Erectus

80   Most cannibalism was not the work of serial killers. Instead it occurred for complex and varied reasons. At an archeological site called Gough's Cave in southwestern England, human bones that are approximately 15,000 years old bear unmistakable signs of cannibalism, like butchering marks and human tooth imprints that suggest even the ends of toes and rib bones were gnawed to get at the last marrow from the bones.

There are also signs of purposeful engraving. There is a zigzag pattern on a human arm bone in Gough's Cave, an indication of ritual. The zigzag design matches patterns on engraved animal bones found in France from the same period, suggesting the design was a common motif. The cannibalized bones showed no evidence that people were harmed. Researchers think they died from natural causes and were then eaten. Cannibalism might have been a way to extend the memory of the dead or honor the dead.

Who Were the Neanderthals? Or What Did Granddad See in Her?

109   Neanderthals and early humans are not the only ones who are fascinated with the handaxe. Recently, in Canada and

some parts of the United States, venues have opened in popular pubs and swingy bars that have competitive indoor axe-throwing contests. There is now an International Axe Throwing Federation which oversees the rules.

113   400,000-year-old fossils of hominins were found inside the Atapuerca cave system in northern Spain. One hundred feet down the cave lies a "pit of bones". The remains of at least 28 ancient humans have been found at the bottom of this 39-foot vertical shaft. DNA evidence of the hominin bones show that these humans, (having the oldest human DNA ever sequenced), pre-dated *Homo sapiens* and were most probably Neanderthals.

No Canvas Needed/ Art in Caves and Art Artifacts

187   The artistic quality of the Stone Age cave art is even more amazing when it is remembered the adverse conditions in which Stone Age painters worked. At Lascaux and 20 other caves in France and Spain, there are hand stencils of mutilated hands. Researchers think that because the thumbs remained on all the hands, the injuries were caused by frostbite.

194   There are indecipherable signs and designs on most of the cave walls along with the paintings. The signs appear alone or in combination with one another. The signs are not actual writing because they don't repeat the way writing does. Researchers think these signs are a code and each has a specific meaning. The presence of the signs suggests that the paintings didn't stand on their own. They needed some extra descriptions or comments.

203   There are theories that "grannies" or older females who have gone through menopause could have been some of the people who helped to paint the images on the caves. The theory is that the grannies were not burdened with raising babies but were older and held the wisdom of their culture. They were

among the people who were tasked to relate the values of their tribe on the walls of the caves.

205   There are numerous theories about the human psyche and the emergence of the "Great Mother Goddess". In these theories, fertility and giving birth equals power. The theories start with the Venus figurines of 40,000 years ago. Analysis focuses on the history of societies in which men now hold power. The conclusions are that the eventual separation of the feminine which has given way to a patriarchal elite has separated humans from the world of nature

The themes of fertility and power fill the history of humanity. Slowly, the theories stress, there was a transition to patriarchal culture that historically goes along with the advancement in weaponry.

Are We There Yet?

208   Try taking the free "Implicit Bias Test" created from Harvard. It is called "Project Implicit" and can be found at implicit.harvard.edu

Generalized biases are not necessarily conscious prejudices and may affect ideas of fairness.

Bibliography and Recommended Reading

Bahn, Paul G., *Cave Art: A Guide to the Decorated Ice Caves of Europe*, Frances Lincoln Limited, 2007

Chua, Amy, *Political Tribes*, Penguin Random House, 2018

Cleyet-Merle, Jean-Jacques, *The Font-De-Gaume Cave*, Editions Du Patrimoine, 2014

Cleyet-Merle, Jean-Jacques, *Musee National de Prehistorie, Les Eyzies-de-Tayac*, Editions de la Reunion des Musees Nationaux, 2007

Curtis, Gregory, *The Cave Painters: Probing the Mysteries of the World's First Artists*, First Anchor Books, 2007

Darwin, Charles, *Origin of the Species; 150th Anniversary Edition*, Market Paperbacks, 2003

Davis, James C., *The Human Story: Our History from the Stone Age to Today,* Harper Perennial, 2004

Delluc, Brigitte and Gilles, Alain Roussot, *Perigord Prehistory*, Editions SudQuest, 2011

Desdemanines-Hugon, Christine, *Stepping Stones*, Yale University Press, 2010

Diamond, Jared, *The World Until Yesterday*, Penguin Group, 2012

Dudley, Robert, *The Drunken Monkey: Why We Drink and Abuse Alcohol*, University of California Press, 2014

Geneste, Jean-Michel, *Lascaux*, Gallimard Decouvertes, 2013

Harari, Yuval Noah, *Sapiens: A Brief History of Humankind*, Harper Collins, 2015

Hawkes, Jacquetta, *The Atlas of Early Man*, Kindersley Limited, 1993

Hendy, Ivy, *Docent Details*, I Street Press, 2013

Linsey, Jennifer, *Jane Goodall, 40 Years at Gombe: A Tribute*, Stewart, Tabor, & Chang, 1999

Lorenz, Konrad, *On Aggression*, Harvard Books, 1963

Rosling, Hans, *Factfulness*, Flatiron Books, 2018

Rudgley, Richard, *The Lost Civilizations of the Stone Age*, Penguin Books, 1999

Rutherford, Adam, *A Brief History of Everyone Who Ever Lived: The Stories of Our Genes*, Workman Publishing, 2017

Stapolsky, Robert M., *Behave: The Biology of Humans at our Best and Worst*, Penguin Press, 2017

Shipman, Pat, *The Invaders: How Humans and their Dogs Drove Neanderthals to Extinction*, Harvard University Press, 2017

Standage, Tom, *The History of the World In 6 Glasses*, Bloomsbury, 2006

Stanley, Steven M., *Children of the Ice Age*, Harmony Books, 1996

Stringer, Chris, and Peter Andrews, *The Complete World of Human Evolution*, Thames & Hudson, 2014

Stott, Rebecca, *Darwin's Ghosts*, Spiegel & Grau, 2012

Wilson, Edward O., *The Meaning of Human Existence*, W.W. Norton, 2014

Wrangham, Richard, *Catching Fire: How Cooking Made Us Human*, Perseus Books, 2010

Suggested Viewing

Great Courses DVDs:

Archaeology: Introduction to the World's Greatest Sites

Between the Rivers: History of Ancient Mesopotamia

Big History

Food, Science, and the Human Body

Major Transitions in Evolution

Rise of Humans

Roots of Human Behavior

Story of Human Language

Miscellaneous DVDs/Recommended Viewing:

Cave of Forgotten Dreams

Iceman

Quest for Fire

Index

Abstract art, 184, 185
Acheulean, 133, 155, 160
African, 10, 26, 27, 34, 35, 43, 53, 54, 70, 75, 84, 134, 143, 144, 153, 155
Altamira, 5, 187, 188, 190, 192
Altruism, 173
Ancestor Worship, 174
animism, 175, 176
Asia, 77, 95, 96, 118, 125, 127, 128, 130, 135, 139, 140, 145, 147, 149, 155, 161, 162, 165
Atapuerca cave system, 216
atlatl, 168
aurochs, 98, 146, 192, 193, 194
Australopithecus, 35, 36, 37, 39, 40, 41, 42, 43, 44, 45, 46, 49, 51, 61, 62, 131, 136, 137, 151, 210, 215
axe, 12, 133, 157, 216
bees, 85, 86, 163
Beliefs, 99, 101, 171, 172, 176
Bill Gates, 127
Blombos cave, 186
bone industry, 202
cannibalism, 41, 80, 81, 180, 215
cave art, 184, 185, 191, 193, 194, 195, 197, 198, 199, 201, 202, 216

Ceremonies, 181, 199
chimpanzees, 11, 33, 34, 35, 40, 41, 57, 65, 74, 214
China, 62, 77, 78, 81, 125
Climate Extinctions, 214
cognitive revolution, 166
compassion, 11, 19, 77, 99, 112, 211
complexity, 9, 144, 166, 170, 182
conformity, 177, 178, 179
cooking, 15, 30, 63, 65, 66, 67, 132, 177
Coprolites, 98
Cro-Magnon, 165, 166, 167, 168, 186, 201, 202
dance, 200
Darwin, 16, 17, 52, 136, 142, 211, 218, 220
Death rituals, 116
Deities, 175
Demons, 176
Denisovans, 8, 124, 125, 126, 127, 128, 129, 130, 131, 134, 135, 140, 141, 164
Division of Labor, 55
Dmanisi, 76, 77
Drunken monkey theory, 151
Eat-a-Bug Cookbook, 68
Equal Partners, 90
evolutionary pressures, 19, 103

fire, 10, 12, 26, 28, 63, 64, 65, 66, 67, 68, 83, 84, 90, 110, 117, 120, 121, 132, 157, 161, 175, 183, 187, 207
fishing platforms, 129
FoxP2, 104
Fruit, 149
gender-based division, 90
Geometric Designs, 194
Gossip, 178
Gough's Cave, 215
grave goods, 102, 173, 174
Green Sahara, 153
handaxe, 12, 79, 80, 84, 167, 215
Hobbits, 118, 119, 120, 121, 122, 123, 135
Holocene, 30
hominid, 4, 21, 83
hominin, 10, 11, 19, 21, 23, 26, 28, 29, 30, 31, 32, 33, 35, 39, 41, 43, 48, 49, 52, 53, 54, 55, 56, 57, 58, 59, 61, 62, 63, 66, 70, 72, 74, 75, 77, 87, 88, 90, 114, 115, 116, 117, 118, 119, 120, 122, 123, 124, 126, 128, 129, 131, 132, 133, 134, 135, 136, 139, 140, 144, 151, 152, 153, 160, 163, 167, 168, 207, 216
Homo erectus, 21, 28, 29, 50, 52, 59, 61, 62, 63, 64, 66, 68, 70, 72, 74, 76, 77, 78, 79, 80, 81, 82, 83, 84, 85, 89, 96, 119, 122, 128, 133, 135, 136, 137, 139, 140, 152, 153, 154, 156, 159, 160, 161, 162, 165, 186, 201, 207
Homo ergaster, 50, 52, 133, 134, 140
Homo floresiensis, 7, 114, 118, 119, 120, 122, 134, 135
Homo habilis, 4, 21, 43, 44, 45, 46, 47, 48, 49, 50, 52, 53, 55, 59, 66, 80, 89, 133, 138, 160, 212, 215
Homo heidelbergensis, 85, 86, 87, 88, 89, 90, 92, 96, 104, 133, 134
Homo naledi, 7, 114, 115, 116, 117, 118, 134, 135, 210
Homo rudolfensis, 49, 50, 134
Homo sapiens, 7, 11, 14, 18, 21, 26, 28, 29, 30, 33, 35, 43, 52, 75, 94, 97, 98, 104, 106, 107, 108, 109, 110, 111, 114, 117, 118, 124, 127, 131, 134, 136, 137, 139, 140, 142, 143, 144, 152, 156, 158, 159, 160, 161, 162, 163, 165, 166, 184, 185, 206, 207, 210, 211, 216
Homo tsaichangensis, 160
honey, 85, 86
Hyoid, 87, 88, 104
Inequality, 91
interbreeding theory, 107

interglacial, 24, 25
Israel, 100, 149, 155, 156, 157, 158, 162
Jane Goodall, 212, 214, 219
Jebel Irhoud, 158, 159
Laetoli Footprint, 215
language, 5, 23, 86, 87, 88, 89, 90, 104, 134, 163, 169, 175, 179, 182, 183
Language Theories, 86
larynx, 87, 88, 103
Lascaux, 5, 192, 193, 194, 195, 196, 216, 219
Liang Bua, 118
Louis Leakey, 4, 43
Lucy, 35, 36, 37, 38, 39, 40, 41, 42, 61
megafauna, 26, 27, 32
migrations, 152, 156, 197
Misliya, 157, 158
Morocco, 158
mortuary practices, 116
Mousterian, 96
music, 200, 201
Natural Selection, 16, 75, 105, 132, 140, 141, 142, 175, 206, 208, 211
Neanderthal, 8, 92, 93, 94, 95, 96, 97, 98, 99, 100, 101, 103, 104, 105, 106, 107, 108, 111, 113, 124, 130, 174, 179, 180, 185, 186, 187
Ochre, 76, 166, 172, 183, 184, 185, 186, 189
Olduvai, 4, 43, 77, 82, 156, 206, 215

Out of Africa, 77, 138, 152, 154, 155
pair bonding, 41, 43, 133
Paleolithic, 12, 14, 15, 59, 76, 100, 103, 131, 146, 147, 149, 165, 182, 201, 214
patriarchal culture, 217
Pleistocene, 118, 214
Portable art, 201, 202
Qafzeh, 156
Rafts, 78, 128
religion, 14, 109, 172, 173, 175
Rising Star, 114, 116, 117
rituals, 116, 117, 172, 181, 182, 199
Shaman, 197
Sharing Primate Characteristics, 33
shells, 12, 65, 78, 79, 166, 183, 186, 201, 202, 203, 211
signs, 19, 43, 50, 76, 78, 117, 120, 134, 172, 192, 194, 195, 215, 216
social identity theory, 210
Sorcerer, 197, 200
speciation, 141, 159
stone knapper, 214
Stone lamps, 187, 189, 200
sweat glands, 71, 75
sweatiest primates, 72
symbols, 13, 194, 195, 196, 205, 210
Taiwan, 160
taxonomy, 52

Terra Amata, 76, 83
Theory of the Hand, 191
trade, 166, 203, 211
Tribalism, 172, 178, 208
tsaichangensis, 160
Turkana Boy, 50, 51
vegetables, 64, 90, 132, 148, 149
Venus figurines, 15, 203, 204, 211, 217
Venus of Willendorf, 204, 205
weapons, 80, 107, 108, 109, 110, 112, 163, 167, 168, 170, 177, 192, 202
zigzag, 79, 186, 215

Made in the USA
Las Vegas, NV
22 June 2021